SCIENCE AND ECCENTRICITY: COLLECTING, WRITING AND PERFORMING SCIENCE FOR EARLY NINETEENTH-CENTURY AUDIENCES

SCIENCE AND CULTURE IN THE NINETEENTH CENTURY

Series Editor: Bernard Lightman

TITLES IN THIS SERIES

FORTHCOMING TITLES

www.pickeringchatto.com/scienceculture

SCIENCE AND ECCENTRICITY: COLLECTING, WRITING AND PERFORMING SCIENCE FOR EARLY NINETEENTH-CENTURY AUDIENCES

BY

Victoria Carroll

LONDON
PICKERING & CHATTO
2008

Published by Pickering & Chatto (Publishers) Limited
21 Bloomsbury Way, London WC1A 2TH

2252 Ridge Road, Brookfield, Vermont 05036-9704, USA

www.pickeringchatto.com

Carroll, Victoria
Science and eccentricity : collecting, writing and performing science for early
nineteenth-century audiences. – (Science and culture in nineteenth-century
Britain)
1. Science – Great Britain – History – 19th century 2. Technical writing – Great
Britain – History – 19th century 3. Eccentrics and eccentricities – Great Britain
– History – 19th century
I. Title
509.4'1'09034

ISBN-13: 9781851969401

This publication is printed on acid-free paper that conforms to the American
National Standard for the Permanence of Paper for Printed Library Materials.

Typeset by Pickering & Chatto (Publishers) Limited
Printed in the United Kingdom by MPG Books Ltd, Bodmin, Cornwall

CONTENTS

ACKNOWLEDGEMENTS

I am indebted first and foremost to Nick Jardine and Jim Secord. Nick first introduced me to the nineteenth-century naturalist Charles Waterton, and it was as a result of this encounter that the idea of a study of science and eccentricity was born. Since then, Nick has continued to be an inspirational mentor and a reliable friend; his encouragement and support have been unfailing. Jim Secord too has been a crucial influence from the very beginning of this project. I would like to thank him sincerely for his guidance, his enthusiasm and his immense generosity, without which I would not have been able to write this book.

Many other people have helped me along the way. I am especially grateful to Katey Anderson, Simon Schaffer, Jon Topham and my editor Bernie Lightman, who, along with Nick Jardine and Jim Secord, have commented extensively on drafts of the whole manuscript. Individual chapters have benefited from the comments of Dan Albert, Sam Alberti, Tim Boon, Ali Boyle, Robert Bud, Alison Conboy, Sophia Davies, John Forrester, Marina Frasca-Spada, Aileen Fyfe, Miranda Gill, James Gregory, Katherine Haddon, Nigel Leask, Peter Mandler, Jack Morrell, Andrew Nahum, Ayesha Nathoo, Ralph O'Connor, Lisa O'Sullivan, Kate Parks, Julia Pitts, Sadiah Qureshi, Rob Ralley, Nicky Reeves, Martin Rudwick, Anne Secord, Rebecca Stott, Mike Taylor and Jenny Ward. Many other colleagues from the Department of History and Philosophy of Science in Cambridge and the Science Museum in London have provided invaluable feedback and support, for which I am immensely grateful. I have benefited greatly from invitations to present work at seminars and conferences: I would particularly like to mention the Cabinet of Natural History, Cambridge, and the 'Nineteenth-century Popular Science: Sites and Experiences' conference, organized by Bernie Lightman and Aileen Fyfe and held at York University, Toronto, in 2004. Earlier versions of Chapter 4 have previously appeared in the book which came out of that conference, A. Fyfe and B. Lightman (eds), *Science in the Marketplace: Nineteenth-Century Sites and Experiences* (Chicago, IL: University of Chicago Press, 2007), and in *Studies in History and Philosophy of Biological and Biomedical Sciences*, 35 (2004), pp. 31–64. A version of Chapter 3 has appeared in *Isis*, 98 (2007), pp. 225–65.

For providing access to archives, images, objects and other information, I would like especially to thank Keith Armstrong, Brian Edginton and staff at Wakefield Museum, the Whipple Library, the Whipple Museum, the University of Cambridge Library, York Central Reference Library, Newcastle City Archives, Stonyhurst College, the Natural History Museum and the Laing Art Gallery.

This book has developed out of my PhD dissertation, which I completed at the Department of History and Philosophy of Science, University of Cambridge. I received funding during the PhD from the Arts and Humanities Research Board, the Isaac Newton Trust, the Raymond and Edith Williamson Fund and Jesus College, Cambridge. I am extremely grateful to the Science Museum for giving me a sabbatical in which to complete the book.

Finally I would like to thank my friends, my family and, especially, my parents, to whom this book is dedicated.

LIST OF ILLUSTRATIONS

INTRODUCTION

During his third expedition to South America, in 1820, the Roman Catholic naturalist Charles Waterton found himself in a rather awkward situation. An avid collector and taxidermist, Waterton had travelled 540 km through the 'wilds of Guiana' with his entourage to obtain a perfect cayman specimen for the museum at his country house back in Yorkshire. Now, faced with the furious beast stuck fast to the end of a rope – it had swallowed a baited hook cast into the Essequibo river – Waterton was at a loss. His helpers were divided: some wanted to kill it with arrows, whilst others preferred to shoot it. Either would have been disastrous. Waterton wanted a perfect specimen, not a mutilated one, and so there was only one thing for it: it had to be taken alive. Grabbing a canoe mast for protection, Waterton crouched down by the bank of the river, holding the mast like a bayonet, and ordered the men to haul the unhappy reptile out of the water. As soon as it was landed, he leapt fearlessly onto its back, turning half round as he vaulted, and grabbed hold of its forelegs, twisting them behind its back to serve as a bridle as the men continued to drag the pair further inland to safety. 'It was the first and last time I was ever on a cayman's back', Waterton later explained. 'Should it be asked, how I managed to keep my seat, I would answer, – I hunted some years with Lord Darlington's fox hounds'.[1] In 1825, Waterton published an account of this and many other adventures as *Wanderings in South America, the North-west of the United States and the Antilles, in the Years 1812, 1816, 1820 & 1824*. Before long, the cayman anecdote had been reprinted in virtually all the newspapers, and the caricature shown in Figure I.1 could be seen in print shop windows across Britain. People flocked from all around to the home of the celebrated naturalist to view his collection of specimens, the unfortunate cayman amongst them.

The history of nineteenth-century British science is replete with anecdotes. They are what lend the period much of its charm. We hear of the young Charles Darwin scavenging around on the banks of the river Cam for rare beetles, greedily stuffing one into his mouth to allow him to catch a third, only to have to spit it out again when it squirts a noxious fluid down his throat.[2] We see Richard Owen and his palaeontological and geological cronies squeezed around a

Figure I.1. *It Was the First and Last Time I Was Ever on a Cayman's Back* (1827), by the well known caricaturist Robert Cruikshank, first appeared in the second edition of *Wanderings in South America*. The print was also available from booksellers and print shops. It was republished in 1836 due to popular demand. Reproduced by kind permission of Jim Secord.

dinner table, toasting the Queen, inside Waterhouse Hawkins's life-size model iguanodon at the Crystal Palace in Sydenham.[3] And we read of Charles Babbage snacking leisurely on Irish whiskey and biscuits only moments after having been hauled out of the seething crater of Vesuvius, his boots destroyed by intense heat, his walking stick consumed by flames.[4] Being concerned with science and eccentricity, this book boasts more than the average share of extraordinary anecdotes. Typically, such anecdotes have been treated by historians of science as illustrative material: as hooks to engage potential readers, as biographical curiosities, or as entertaining decorations to more 'serious' arguments. This book, by contrast, considers them as phenomena worthy of attention in their own right.

One premise of this book is that eccentricity is an historically and culturally specific phenomenon, and it is perhaps worth pointing out that this is far from self-evident. Previous analyses of eccentricity, from prosopographical collections of 'eccentric biography', which date back to the turn of the nineteenth century, to more recent psychological studies such as David Weeks's and Jamie James's *Eccentrics* (1995), have treated eccentricity as something like a human universal, something which has endured 'in All Ages and Countries'.[5] This book, building on more recent scholarship, argues that a British discourse of eccentricity (which sometimes incorporated figures from history) became established only in the early nineteenth century, within a specific, formative social and cultural context.[6] The book asks: How and why did a discourse of eccentricity emerge and flourish at this time? Why was science a focus for eccentricity? And what can the study of 'eccentric' naturalists, collectors and natural philosophers add to our understanding of the pursuit and communication of natural knowledge in a period generally understood by historians to have produced many of the defining features of modern science?

Important changes were taking place in the natural sciences in Britain in the first half of the nineteenth century. Very broadly speaking, this was the period in which many branches of science began to undergo a gradual process of specialization. As the discussion of new scientific research began to occur in dedicated journals rather than in the literary quarterly magazines, which catered to a highbrow but general audience, the foundation of specialist learned societies – the Linnaean Society in 1788, the Geological Society of London in 1807, the Zoological Society in of London in 1826, the Royal Astronomical Society in 1831 and the Chemical Society of London in 1841 – provided institutional cohesion for practitioners in what were coming to be seen as discrete scientific disciplines. This was also the period in which some (though by no means all) British men of science, lagging several decades behind their French and German counterparts, began to see themselves as professionals rather than as gentlemen enthusiasts. Following Charles Babbage's and David Brewster's bitter complaints about the 'decline of science in England' compared with the situation on the Continent, the British Association for the Advancement of Science, founded in 1831, began to represent the career interests of scientific practitioners, who would gradually start to be known as 'scientists' after 1833, when William Whewell coined the word. To a limited extent, new opportunities for paid employment in science were created: as lecturers in universities and in provincial literary and philosophical societies and mechanics' institutes; as naturalists and astronomers on board the exploratory naval voyages which underpinned the expansion of the British Empire; and through government-sponsored projects such as the Geological Survey, which was founded in 1835. While France and Germany led the way in university-based research and the provision of formal scientific training, in Britain organized science would become a regular feature of the universities only in the second half of the century. Nevertheless, in Britain as well as on the Continent, science gradually became open to careerists as well as gentlemen.[7]

This familiar historiographical narrative, according to which the modern, professional scientist emerged and gained strength over the course of the nineteenth century, is a simplified one, and it has been challenged from various directions. Whilst it cannot be denied that significant changes, as outlined above, did indeed take place, they did so haphazardly and, more importantly for this study, as part of a bigger picture in which other developments were taking place at the same time. For example, in the early nineteenth century, for the first time, scientific knowledge became available for mass consumption by non-specialist audiences from across the social spectrum through an explosion of lectures, exhibitions, books, libraries, magazines and clubs. Non-specialists engaged in scientific leisure pursuits, such as natural history and collecting, more than ever before. Science was made 'popular', with all the ambiguities that word entails.[8] Conversely, activities now considered 'unscientific' – mesmerism, phrenology, psychical research – continued to be practised with enthusiasm by established members of the scientific elite.[9] Like 'popular' and 'marginal' science, and like the contributions of 'invisible' workers, such as technicians, artisan collectors, educators and women, science as practised by supposedly 'eccentric' individuals has been prone to exclusion from the historiography of the sciences over the years.[10] The present study draws inspiration from the many histories of the popular, the marginal and the invisible which, over the last decade or so, have greatly enriched our understanding of the range and diversity of scientific culture in the period. My aim in this book, however, is not so much to recuperate 'eccentric' practitioners of the sciences into 'mainstream' historiographical narratives – I do not wish to argue that their activities were just as much a part of institutionally sanctioned science in their day as those activities now considered to be 'properly' scientific – rather, through focusing on individuals who *really were* considered to be marginal in their own time, my aim is precisely to explore the significance of marginality within (and without) early nineteenth-century scientific culture.

Through considering concrete historical cases in which boundaries were contested and equivocal sentiments were expressed, I hope also to reveal something more of the hidden structures underlying ordinary, uncontested scientific practice. In anthropology it is generally understood that marginal figures, far from being best ignored, are especially useful when it comes to understanding a culture. Barbara Babcock states the case particularly clearly in *The Reversible World* (1978), in a discussion of 'symbolic inversion'. She defines symbolic inversion as 'any act of expressive behaviour which inverts, contradicts, abrogates, or in some fashion presents an alternative to commonly held cultural codes, values and norms be they linguistic, literary or artistic, religious, social and political', to which we can add 'scientific'.[11] Other scholars, such as Peter Stallybrass and Allon White, have called this phenomenon 'transgression', a term which I will use.[12] Babcock argues that, 'far from being a residual category of experience', symbolic inversion 'is its very opposite. What is socially peripheral is often symbolically central, and if we ignore or minimize inversion and other forms of cultural negation, we often fail to understand the dynamics of symbolic processes generally'.[13] Individuals were labelled 'eccentric' in the nineteenth century, I will argue, when they were seen to transgress boundaries. In this period, we have seen, practitioners of the sciences were especially preoccupied with negotiating boundaries: between the various disciplines, between 'professional' and 'amateur', between science and non-science, between 'elite' and 'popular'. 'Eccentric', transgressive scientific practitioners tended to be marginalized socially; yet, often, they provoked a disproportionate degree of response and reaction. At the symbolic level, 'eccentric', marginal figures represented concerns that were of central importance to scientific and local communities.

The book begins with a discussion of how a discourse of eccentricity was constructed in Britain around the turn of the nineteenth century, and how that discourse developed in a variety of cultural contexts over the following decades. The net is deliberately cast wide in the first instance: poetry, fiction, history, biography, philosophy and news are trawled for evidence of how people spoke, wrote and thought about eccentricity. The aim of the first chapter is to build up a multilayered, contemporary definition. Rather than ask, 'How should I define eccentricity?' I ask, 'How did people define it at the time?' Eccentricity is examined in connection with a wide variety of themes including astronomical symbolism, the 'Spirit of the Age', prophecy, classification, gender, genius, the popular press, madness and classification. Running through the chapter, and indeed those that follow, is a concern with boundaries: people, objects, events and natural phenomena were labelled 'eccentric' in this period, I argue, if they were perceived to transgress the boundaries which ordered social and cultural life.

Having established a broad cultural definition of eccentricity, as it was understood in the early nineteenth century, the relationship between eccentricity and science in the period is investigated in depth through a series of closely focused case studies. Each of the three remaining chapters concerns a single individual who was engaged in the pursuit of natural knowledge. However, the focus of each case study is thematic rather than biographical. The first criterion by which the three individuals – the natural philosopher William Martin, the fossilist Thomas Hawkins, and the naturalist Charles Waterton – were selected is that each was explicitly labelled 'eccentric' by contemporaries. None of them, however, publicly professed to be 'eccentric' themselves, which leads me to an important methodological point. 'Eccentric' was, by and large, a label which was applied to individuals by others. In order to understand eccentricity, therefore, it is necessary to understand the reasonings and behaviours of those others. These 'audiences' for eccentricity were, through their interactions with and responses to 'eccentric' individuals, crucially implicated in the construction of those individuals' eccentric identities.

Divergent varieties of audience-oriented criticism have in recent years begun increasingly to find applications in the history of science. A surge of interest in the wider public engagement with science has resulted in the near demolition of simplistic diffusion models of science communication, according to which scientific knowledge drifts passively outwards from experts and is absorbed by a monolithic lay audience.[14] A new emphasis on the material means of communication, driven by developments in book history, has shifted awareness away from *texts*, towards the production, distribution and consumption of scientific *books*; this awareness has extended outwards to encompass scientific lectures, shows, demonstrations and collections.[15] But perhaps the most significant innovation has been the new attention paid to varieties of reading, in the broad sense. Media and literary studies have been influential in this connection. In the 1970s, Stuart Hall's seminal paper arguing for the asymmetry of codes by which televisual messages are 'encoded' and 'decoded' urged students of the mass media to look to audiences rather than programme content in their efforts to explain media effects.[16] Parallel movements in literary studies resulted in a new emphasis on readers, both as they are inscribed within literary texts, and as they exist outside of texts as interpreters.[17] In history of science, more recently, works such as Jim Secord's *Victorian Sensation: The Extraordinary Publication, Reception, and Secret Authorship of Vestiges of the Natural History of Creation* (2000) have demonstrated that, by reconstructing the contexts in which scientific books were circulated, read, reproduced, reviewed and talked about, and by examining the testimonies of individual readers, it is possible to recover the meanings particular scientific books had for different communities of readers.[18] The case studies in this book draw on these interconnected literatures in constructing a framework within which to situate the various meanings which nineteenth-century 'eccentric' persons, practices and things had for different reader/audience groups.

The case studies deal with audience responses to a wide range of authored productions, from taxidermic specimens and inventions, to illustrations, performances and printed texts. The responses themselves take a variety of forms. News reports, visiting accounts, personal reminiscences, book reviews, cartoons, polemical pamphlets, private correspondence, biography – these are just some of them. Each type of source has its advantages and its limitations. Reminiscences, for example, can provide personal insights into the characters of 'eccentric' individuals, but when they are written long after the event, they are prone to omissions and inaccuracies; more importantly, they inevitably frame past events from the point of view of another age. Book reviews give an indication of how books were perceived by contemporary readers, but they are necessarily shaped by generic constraints. Published visiting accounts are helpful in recovering something of the lived experiences of real people as they encountered supposedly eccentric individuals in the flesh, but they cannot be treated as direct portals into the minds of their authors: they too are shaped by the expectations of editors and readers. Throughout the case studies, therefore, I aim not to take such sources too much 'at face value'. Rather, by considering individuals' responses in relation to broader 'horizons of expectations' – based, for example, upon conventions of genre, custom and tradition – I aim to describe the patterns of reception within which eccentricity was defined.

Chapter 2 argues explicitly for the active roles of audiences in the creation of eccentric identities. It does this through analysing responses to the lectures and demonstrations of the self-taught, anti-Newtonian natural philosopher and prophet, William Martin (1772–1851). Martin published a pamphlet entitled *A New System of Natural Philosophy on the Principle of Perpetual Motion* in 1821.[19] In 1827, at God's command, he began to lecture on his philosophy in pubs and other venues in and around his hometown, Wallsend, near Newcastle upon Tyne in the north of England. These performances drew large, often rather raucous audiences, who actively engaged with the Martinian 'New System' by haranguing the philosopher and publishing ballads and broadsides in response to him. The chapter begins by recreating Martin's lectures, including the important contributions of his self-styled 'Disciples' and adversaries. It then examines Martin's new system of philosophy in more depth, locating his approach with respect to traditions of anti-Newtoniansm and millenarian prophecy. Martin's beliefs were highly unorthodox but they had precedents; he was labelled 'eccentric' but his views would not have appeared utterly nonsensical to contemporaries (as they perhaps do today). Indeed, I argue that it was precisely because Martin existed at the margins of Newcastle's rich intellectual and scientific culture, and not entirely within or outside it, that he could be symbolically so powerful. Drawing on traditional forms of custom and carnivalesque expression, Martin's Disciples and opponents used Martinian performance as a means of addressing serious and highly politicized concerns: about religious tolerance, freedom of expression, political reform, and the 'improvement' of the masses through the provision of 'useful knowledge'.

Writing and publishing were central to Martin's eccentric identity. Chapter 3 turns to the writings of the fossil collector Thomas Hawkins (1810–89) in order to deconstruct the notion of literary eccentricity further. The study of geology was a highly fashionable activity in the early nineteenth century, popular with both men and women. From around the 1830s, however, an increasingly esoteric group of practitioners associated with the Geological Society of London began intellectually and institutionally to define themselves as the elite of 'scientific' geology: one strategy was to distance themselves from 'scriptural geologists', who, they claimed, placed too much emphasis on the word of God as a source for geological knowledge. Hawkins was keen to be associated with the scientific elite. He produced two giant volumes describing his exceptional collection of fossil reptiles – *Memoirs of Ichthyosauri and Plesiosauri* (1834) and *The Book of the Great Sea-Dragons* (1840), but critics complained both were incomprehensible due to the 'eccentric style in which [they were] written and put together'.[20] Through a combination of close reading and reception analysis, the chapter argues that Hawkins's books were deemed eccentric because they challenged important boundaries: they were seen to undermine the generic conventions which typically governed the production and reception of scientific works. Particularly controversial were, first, the central place he gave to biblical accounts of Creation in his palaeontological writings and, second, his penchant for writing in an obscure, visionary style more commonly associated with certain kinds of epic poetry. By introducing the notion of genre, the chapter sheds further light on the role of audience responses, in this case reader responses, in the construction of eccentric scientific reputations. Genre binds authors and readers together through the generation of expectations, and it was when expectations were unmet that works were judged 'eccentric'.

The theme of expectation is also at the heart of Chapter 4, which explores the roles of generic frameworks and conventions of visiting in shaping people's experiences of 'eccentric' objects on display. It focuses on the responses of visitors to Walton Hall, home to Charles Waterton (1782–1865), suggesting that visitors considered the 'eccentric' naturalist himself to be one of the key exhibits there. Working from first-hand visiting accounts, drawn from memoirs, travel books and periodicals, the chapter explores how visitors interconnected objects, specimens, stories and images to make their experiences of Walton Hall meaningful. In particular, the influence of Waterton's *Wanderings in South America* on visitors' interpretations of his taxidermic specimens is explored.[21] Charles Waterton was famous, and the chapter locates Waterton and his visitors in the context of the rise of celebrity culture in the first half of the nineteenth century. Visitors recorded their experiences of Waterton and his home as anecdotes, to be told and retold to future generations of visitors and readers, and the chapter argues that these anecdotes were crucial to the construction of Waterton's status as a celebrity figure, and to the propagation of his historical reputation as one of England's great eccentrics. I conclude with some reflections on the historical legacies of Martin, Hawkins and Waterton to date.

The public image of the scientist is now a key research topic in science studies. The representation of present-day scientists in the media, for example, is a central concern within the rapidly growing field of science communication. The portrayal of real and fictional scientists in literature, drama and film is a major focus of interdisciplinary research conducted under the banner of 'science and literature'.[22] In history of science more narrowly defined, the processes by which scientific identities and reputations are constructed and transformed have increasingly attracted the attention of scholars in recent years. While certain recurrent tropes – the scientist as genius, for example – have been subjected to incisive historical and critical deconstruction, the scientist as eccentric has not. Indeed, such research has until now tended to focus on individuals who enjoyed exceptionally 'successful' careers and who today hold coveted places in the historic halls of scientific fame.[23] By dealing with the public identities and historical reputations of marginal, contested, 'eccentric' figures in the early nineteenth century, this study aims to redress the balance, providing a new perspective on cultures of natural knowledge in a period which not only prepared the ground for the emergence of the modern, specialist, professional scientist but also, simultaneously, supported a thriving population of highly visible, 'eccentric' scientific practitioners. These developments were inextricably linked and each can, I suggest, be better understood in light of the other.

1 DEFINING ECCENTRICITY IN EARLY NINETEENTH-CENTURY BRITAIN

In the 1869 preface to a *Book of Wonderful Characters; Memoirs and Anecdotes of Remarkable and Eccentric Persons in All Ages and Countries*, an unnamed editor made two claims for the significance of the work. First, the biographies of individuals 'possessing an eccentricity of character' had, he wrote, 'been in all times eagerly sought after by the curious inquirer into human nature'. Eccentricity was a perennially fascinating subject. Second, though, 'a great change ... in the manners and customs of the people of England' had in recent years rendered the subject of eccentricity yet more acutely interesting: 'We have nearly lost all, and are daily losing what little remains of, our individuality', he wrote with alarm; 'all people and all places seem now to be alike'.[1] This mid-Victorian writer was not alone in fearing himself to be living in an age of decline, as far as eccentricity was concerned. A decade previously, in what is now probably the most frequently cited nineteenth-century dictum on eccentricity, the philosopher John Stuart Mill warned:

> Eccentricity has always abounded when and where strength of character has abounded; and the amount of eccentricity in a society has generally been proportional to the amount of genius, mental vigour, and moral courage it contained. That so few now dare to be eccentric marks the chief danger of the time.[2]

Mill's primary concern in his treatise *On Liberty* (1859) was the extent to which power could rightly be exerted over members of a civilized society; the demise of eccentricity, in his opinion, was a symptom of the tyranny of public opinion being allowed to interfere unjustifiably with the liberty of individuals. The editor of the *Book of Wonderful Characters*, taking a different angle, blamed the material fruits of the modern industrial age: the railways, the steam press, technologies of mass production. The latest scientific discoveries, he observed, once the monopoly of a handful of natural philosophers, were now, thanks the expansion of the popular press, 'the common property of the multitude'; works of literary genius were 'so cheap as to be found on the labourer's shelf'; recent advances in engraving and casting meant that reproductions of great artworks were 'the refined household treasures of millions'. The dramatic increase in personal mobility occasioned by the railways coupled with a technologically driven democratization of natural knowledge and art had, in this author's eyes, culminated in the homogenization of English culture and the death of eccentricity: 'the tendency of the present day, in England', he wrote, 'is directly opposed to the spirit of individual exclusiveness which, as the great encourager of eccentricity of character, once prevailed over all the country'.[3]

These are just two opinions amongst many, but they serve well enough to exemplify a mid-Victorian trend for viewing the recent past as a heyday of eccentricity, the passing of which was to be lamented.[4] Mill and the anonymous editor of the *Book of Wonderful Characters* hazily identified eccentricity as a phenomenon which had a special significance for early nineteenth-century British culture. Recently, cultural historians have just begun to explore this significance more systematically. Brian Cowlishaw's doctoral dissertation, 'A Genealogy of Ecentricity' (1998), represents the first attempt to chart the emergence and historical development of eccentricity as an actors' category in Britain. Paul Langford, in *Englishness Identified: Manners and Character, 1650–1850* (2000), examines eccentricity as one of a number of traits commonly felt to be constitutive of the English national character. Miranda Gill's dissertation, 'Eccentricity and the Cultural Imagination in Nineteenth-century France' (2004), explores the cultural meanings of eccentricity in a diverse array of French contexts, from alienism, to dandyism, to the 'demi-monde'. Most recently, James Gregory has examined eccentricity at the local level, focusing on nineteenth-century eccentric characters in the north-east of England.[5] By way of preparation for the case studies that follow, this chapter builds on recent scholarship to explore further how a discourse of eccentricity emerged and developed in early nineteenth-century Britain.

From the start, eccentricity was tied to science and mathematics. The word 'eccentric' is in its origins a geometric term: it means a circle that is not concentric with another circle. The earliest usage in English given by the *Oxford English Dictionary* is from *The Castle of Knowledge* (1556), by the Welsh mathematician and physician Robert Recorde: 'These two circles ... are eccentrike, for that they haue not one common centre'.[6] In astronomy, 'eccentric' has been used to designate orbits which do not have the Sun – or in the Copernican system, the Earth – at their centres. Comets have historically been called 'eccentrics' on account of their non-circular trajectories, and it is from this association that figurative uses of the word 'eccentric' are derived. People were first called 'eccentric' if they were seen to be like comets: 'He shines excentric, like a comet's blaze!', wrote Richard Savage in 1728.[7]

Until the mid seventeenth century it was generally believed that comets roamed the cosmos as celestial 'wanderers', in contrast to the planets, which

moved in closed orbits. From the 1660s, some astronomers and natural philosophers began to argue that comets were permanent members of the solar system,
rather than anomalous, transient visitors, and that they could be described by
natural laws.[8] In the literary domain, however, comets continued to be associated with irrationality and lawlessness: people were called 'eccentric' if their
behaviour seemed irregular and unbounded, if their lives seemed to follow an
extraordinary course. John Dryden, for example, wrote in 1672 of 'A soul too
fiery, and too great to guide:/ He moves excentrique, like a wandring Star:/
Whose Motion's just, though 'tis not regular'.[9]

   In its early applications, eccentricity was primarily associated with celebrated
individuals, generally men, who appeared unaffected by the laws of nature which
ordinarily limited the possibilities for human achievement. 'Eccentric' men
were bold, original and innovative; they soared above their contemporaries and
achieved great success in their public careers. The astronomical origins of the
metaphor were nearly always made explicit. For example in a pamphlet called
Shaftsbury's Farewel: Or the New Association [*sic*], produced anonymously in
1683 upon the death of the First Earl of Shaftesbury, we find: 'So High as thine
no Orb of Fire can rowl,/ The Brightest, yet the Most Excentrick Soul'.[10] This
usage was still common in the mid eighteenth century. In 1743, for example,
Aaron Hill emphasized the comet's fire, height and brightness, in writing of men
of exceptional military talent:

> Caesars, sometimes, and sometimes Marlbro's rise:
> Comets! That sweep new Tracks, and fright the Skies!
> Not to be measur'd, These, by War's known Laws:
> Form'd, for excentric Fame, and learn'd Applause!
> No Gen'ral System circumscribes their Ways.
> They move, un-rival'd: and were born, to blaze![11]

In June 1787, however, a correspondent of *The Times* could define eccentricity
simply as 'a departure from the general conduct of society'.[12] The metaphor had
become common enough that its astronomical origins no longer needed to be
stated explicitly. And, by the turn of the nineteenth century, eccentricity had
gradually acquired a broader set of cultural meanings. Raymond Williams has
characterized the process by which the meanings of words change historically as
'a problem of *vocabulary*, in two senses: the available and developing meanings
of known words, which needed to be set down; and the explicit but as often
implicit connections which people were making, in … particular formations of
meaning – ways not only of discussing but of seeing many of our central experiences'.[13] The remainder of this chapter explores some of the connections which
people were making during the early nineteenth century, as eccentricity ceased
to be an occasional poetical figure, and became established as a 'keyword', to use

Williams's term, for describing forms of human life, as well as events, abstract entities and man-made and natural objects.

The first section links eccentricity to early nineteenth-century commentaries on the 'Spirit of the Age'. Attempts by high-profile cultural commentators, such as John Stuart Mill and William Hazlitt, to capture the essence of their age and to define it through its representative characters gave rise, I suggest, to an idea of belonging to the present; this in turn drew attention to 'boundary' characters who seemed out of joint with their time. These characters were labelled 'eccentric'. Some were trapped in the past, welded to the outmoded ideas and traditions of bygone ages. Others were ahead of their time. Prophetic, visionary, comet-like figures, these eccentric 'men of the future' could, by virtue of their capacity to think differently, be looked to lead the masses beyond the boundaries of the present and into the future. This prophetic connection will recur in Chapters 2 and 3.

Biographical sketches of remarkable individuals also feature in the second section. It is well known that biography experienced a surge in popularity in the early nineteenth century.[14] Around 1800, a new biographical genre, 'eccentric biography', emerged, and it flourished in the decades that followed. The early nineteenth century is widely acknowledged to have been a crucial period in the history of the British publishing industry, and eccentric biography was just one response to the rapidly growing demand for accessible, affordable printed material.[15] Composed largely of rehashed biographical anecdotes and portraits, the genre catered to the demands of new mass reading audiences, whilst allowing a new breed of professional hack to create 'original' books quickly, using 'scissors-and-paste' editorial techniques. Eccentric biography helped to popularise the concept of eccentricity as a way of dealing with individuals who, like Mill's and Hazlitt's 'men of the future', appeared to challenge boundaries. Moreover, the genre propagated a 'canon' of eccentric characters, and it helped to establish the anecdote as eccentricity's dominant literary form.

The relationship between eccentricity and boundaries is explored in greater depth in the third section. The subjects of eccentric biography were immensely varied, yet the giants, dwarfs, hermaphrodites, misers, hermits, masculine women, prophets and people trapped in the past, who recur time and time again in successive volumes, were systematically portrayed as undermining important classificatory boundaries: between rich and poor, human and non-human, mad and sane, masculine and feminine, then and now. The issue of gender is explored further in the penultimate section. Reaching beyond the limits of eccentric biography, contemporary novels are probed for contemporary attitudes towards gendered eccentricity. One recurrent trope is the eccentric 'masculine' woman. Another is the 'eccentric' man who was felt to be lacking in typically masculine traits. Gendered discourses of eccentricity drew freely on the astronomical metaphor, illuminating the activities of men and women who appeared to move outside their proper spheres. An alternative, equally gendered, vision of eccentricity – eccentricity as lawlessness – is explored in the final section. The novels of Frances Burney, for example, are full of 'eccentric', flighty, comet-like women whose actions are governed by passion rather than reason. In men, too, irrationality and lawlessness were associated with eccentricity but they were also, sometimes simultaneously, associated with 'genius'. The relationship between discourses of eccentricity and of genius are explored further in this section.

Eccentricity, Prophecy and the 'Spirit of the Age'

In his poem of 1743, cited above, Aaron Hill wrote of 'excentric', comet-like persons as objects that 'fright the Skies!' Since ancient times, comets had been understood as signs from which future states of affairs could be prophesied: comets could signify the advent of anything from a change of weather to a natural disaster, from an event of political importance to the end of the world.[16] Although comets could sometimes be auspicious, signifying, for instance, the birth of a monarch, they were usually observed with trepidation. While cometary prognostication had lost much of its prestige amongst the scientific elite by the late seventeenth century, the notion that comets could affect life on Earth through physical means persisted – Newton, for example, argued that comets periodically restored the Earth's fertility.[17] Meanwhile, cometary divination continued to be practised by astrologers and almanac compilers, and remained popular amongst the lower classes.[18] Hill's 'excentric' heroes were like comets that 'fright the Skies' because they brought about great political upheaval: as agents of change they inevitably instilled fear. In the nineteenth century eccentricity continued to be associated with prophecy, but the way in which this association was understood was somewhat different. The nineteenth-century figure of the eccentric prophet was constructed in connection not with practices of cometary prognostication, but with ideas about how individuals related (or did not relate) to the age in which they lived. 'Eccentric' figures were boundary figures, seen to communicate between past, present and future. In order to explain why eccentricity may have taken on these new associations at this time, I turn briefly to the philosopher and political economist John Stuart Mill and the essayist William Hazlitt, and their writings on eccentricity and the spirit of the age.

When, in 1831, Mill published anonymously a series of essays in *The Examiner* on 'The Spirit of the Age', his choice of subject was not an especially original one.[19] Towards the end of the preceding year an anonymous correspondent of *Blackwood's Magazine* had observed, 'That which, in the slang of faction, is called the Spirit of the Age, absorbs, at present, the attention of the world.'[20] Mill's aim was to reflect, as others had recently, on 'what the spirit of the age really is; and how or wherein it differs from the spirit of any other age'.[21] His novel conclusion was that the defining feature of the age was precisely its preoccupation with its status *as an age*. 'The idea of comparing one's own age with former ages, or with our notion of those which are yet to come, had occurred to philosophers', he wrote, 'but it never before was itself the dominant idea of any age.'[22]

That such an age should have fostered a fascination with characters who did not appear fully to belong to their time is, in light of Mill's assessment, unsurprising. For the age of the spirit of the age is by definition an age of comparison. People living in such an age and seeking to understand it must necessarily draw boundaries between past and present, present and future. And upon scrutinizing these boundaries they are bound to find anomalies. For, while the age of the spirit of the age is defined temporally, it is not defined by rigid calendar time; rather, its margins are discerned through reflections and judgements about 'what the spirit of the age really is', and such reflections and judgements can always be challenged. In practice, they were challenged by individuals appearing to belong to, or partake of the spirit of, the past or the future.

One way in which early nineteenth-century writers sought to define the age in which they lived was through the lives of its exemplary figures. In 1825, for example, William Hazlitt published a collection of biographies entitled *The Spirit of the Age: Or Contemporary Portraits*. Paradoxically, only a fraction of Hazlitt's 'contemporary portraits' were portrayed as men of the present. He depicted the novelist Walter Scott, for example, as a man of the past: 'if you take the universe, and divide it into two parts, he knows all that it *has been*; all that it *is to be* is nothing to him'.[23] In his romances, Scott described the manners, characters, superstitions, dialects, costumes, religion and politics of the sixteenth and seventeenth centuries, and through doing so he became 'the most popular writer of the age'. The problem with Scott, Hazlitt argued, was that his politics also belonged to the past: 'Our historical novelist firmly thinks that nothing *is* but what *has been*, that the moral world stands still ... and that we can never get beyond the point where we actually are without utter destruction.'[24]

Hazlitt subjected men of the present to equally incisive criticism. 'Mr. Wordsworth's genius is a pure emanation of the Spirit of the Age', he wrote equivocally. 'Had he lived in any other period of the world he would never have been heard of.'[25] Wordsworth's style was 'one one of the innovations of the time'. 'All the traditions of learning, all the superstitions of the age, are obliterated and effaced.'[26] And yet, being rooted in the present – and being geographically specific too: Hazlitt writes of 'any other *period of the world*' – it had serious shortcomings. Of the *Excursion*, for example, Hazlitt wrote: 'It affects a system without having any intelligible clue to one, and, instead of unfolding a principle in various and striking lights, repeats the same conclusions till they become flat and insipid.'[27]

One of the defining features of commentaries on the spirit of the age was a dissatisfaction with the present, coupled with an enthusiasm for the future. According to Hazlitt, for example, the muse of the present, as embodied in Wordsworth's poetry, shattered the old aesthetic order but offered little by way of replacement. Similar sentiments were expressed by Mill in connection with political establishments of the present. His explanation for the contemporary preoccupation with the spirit of the age was that nineteenth-century Britons realized themselves to be living in an age of transition – 'Mankind have outgrown old institutions and old doctrines, and have not yet acquired new ones', he wrote provocatively.[28] Looking around him, he saw, like Hazlitt, 'men of the past' commanding all men to adhere to the failing institutions of tradition, and 'men of the present' urging the uninstructed multitude make their own ill-informed judgements for themselves. What society needed, Mill argued, were men of the future – prophetic figures, capable of ideas other than those of their age, to guide the multitude in a new and different way.[29]

Hazlitt's biographical volume offered a number of suggestions for potential men of the future. The most prominent was the philosopher and political reformer Jeremy Bentham, a man who had 'offered constitutions for the New World, and legislated for future times'.[30] Bentham devoted himself to abstract philosophical pursuits, producing results which, rather than being rooted in the present, could be generalized for all ages: 'his reasonings, if true at all, are true everywhere alike'. In general habits and in all but his professional pursuits Bentham was 'a mere child'.[31] Neglectful of his personal appearance and averse to company, he concerned himself only with his work: 'He regards the people about him no more than the flies of a summer', Hazlitt explained. 'He mediates the coming age.'[32] Bentham's personal habits, as Hazlitt described them, would have been highly eccentric in the eyes of his contemporaries. Significantly, however, Hazlitt played up these eccentricities, presenting them as signs of Bentham's visionary qualities. After all, it was only because he acted differently from the mass that the philosopher could be expected to lead the nation into the future.

Several decades later, Mill returned to the still-pressing issues of social and political stagnation:

> It does seem, however, that when the opinions of masses of merely average men are everywhere become or becoming the dominant power, the counterpoise and corrective to that tendency would be, the more and more pronounced individuality of those who stand on the higher eminences of thought. It is in these circumstances most especially, that exceptional individuals, instead of being deterred, should be encouraged in acting differently from the mass.[33]

As we saw above, the term which, in 1859, Mill used to describe such non-conformity was 'eccentricity':

Precisely because the tyranny of opinion is such as to make eccentricity a reproach, it
is desirable, in order to break through that tyranny, that people should be eccentric ...
That so few now dare to be eccentric marks the chief danger of the time.[34]

Mill's 'eccentric' men who thought differently, who held ideas other than the
prevalent ideas of the age, were neither 'men of the past' nor 'men of the present',
they were men of the future, secular prophets, spirits of the age to whom alternatives were, in some sense, revealed.

Hazlitt's book of 1825 aimed to represent the spirit of the age through the
lives of its public figures. While the idea of the spirit of the age was new in this
period, collective biography had been a recognized genre of history writing since
ancient times. A new variant of this genre, 'eccentric biography', emerged in
the early nineteenth century. Eccentric biography was, in a sense, a mirror to
Hazlitt's *The Spirit of the Age*, for while Hazlitt was manifestly concerned with
figures who were representative of the age, eccentric biography focused on individuals who did not fit in. And yet the relationship between these two different
kinds of book was not quite so simple. Alongside 'men of the present', Hazlitt
also included in his volume men who seemed to belong to the past and the
future. Conversely, though the subjects of eccentric biography were represented
as essentially marginal figures, they were at the same time felt to reveal underlying truths about the age in which they lived. In both cases, the 'spirit' of the early
nineteenth century was, paradoxically, illuminated through analysis of the lives
of boundary figures, of individuals who flouted the conventions of the times.

Eccentric Biography: The Emergence of a Genre

The nineteenth century has been called an 'age of biography', and it is well known
that written accounts of people's lives experienced a massive boom in popularity
after around 1800: from multi-volume life-and-letters to the brief biographical
notices which became a staple of the popular periodical press; from 'private' letters and diaries to prosopographies of literary talents, good wives, ecclesiastics,
booksellers and virtually every other class of person imaginable.[35] This development was keenly observed by contemporary commentators: 'The present age
has discovered a desire, or rather a rage, for literary anecdote and private history', wrote Walter Scott in his unfinished autobiography in 1808; in 1863 an
anonymous reviewer of Samuel Smiles's *Industrial Biography: Ironworkers and
Toolmakers* echoed, 'This is the age of biography ... there never was a time when
personal memoirs or biographies proper have been so much studied as in the
present century.'[36]

Historians of life writing have invoked a variety of social and cultural factors
to explain the ascendancy of biography. The rise of what are now thought of as
'Romantic' conceptions of genius, it is argued, stimulated interest in the private

lives and personalities of authors, thus creating a fashion for literary biography.[37] The growing preoccupation with personal fame – fuelled, Leo Braudy argues, by 'the unprecedented growth of urban population, the expansion of literacy, the introduction of cheap methods of printing and engraving, the extension of the political franchise, and the revolutionary overthrow of monarchical authority' – led to a greater demand for insights into the lives of celebrities.[38] Biography had always been considered a branch of history, for it provided a window onto past worlds and thus helped the reader to understand past civilizations, but it also served a second function of helping readers to understand human emotion and character.[39] Richard Altick has argued that in the nineteenth century it came to seem as if biography was history itself.[40] The study of abstract historical laws gave way to the study of heroic individuals who shaped human destiny: 'The History of the world is but the Biography of great men', wrote Thomas Carlyle in 1841.[41]

Notions of 'character', too, were centrally important in this period, as Stefan Collini has shown.[42] In the nineteenth century gentlemanliness was redefined in terms of individual character, an essentially moral quality which could be cultivated by members of the lower and middling orders as well as by members of the aristocracy. 'You may admire men of intellect; but something more is necessary before you will trust them', Smiles explained in *Self Help; With Illustrations of Character and Conduct* (1859).[43] Attention to character meant attention to personal appearances, manners and attitudes – the 'trivial things', Smiles called them, of daily life. In an age obsessed with genius and celebrity, individuality and character, biography flourished. And, as the epitome of the individual, the eccentric character was an ideal biographical subject. The rise of a discourse of eccentricity can thus be viewed as part and parcel of the rise of 'the individual' in the nineteenth century, especially as mediated through life-writing.

Biography was a highly heterogeneous genre in the nineteenth century, and one variety, which has received less scholarly attention than others, was 'eccentric biography'.[44] Eccentric biography emerged around the turn of the century and has endured, more or less, to this day. An early example is the anonymously compiled *Eccentric Biography; Or, Sketches of Remarkable Characters, Ancient and Modern, including Potentates, Statesmen, Divines, Historians, Naval and Military Heroes, Philosophers, Lawyers, Impostors, Poets, Painters, Players, Dramatic Writers, Misers, &c. &c. &c. The Whole Alphabetically Arranged; and Forming a Pleasing Delineation of the Singularity, Whim, Folly, Caprice, &c. &c. of the Human Mind.*[45] This work was published in London in 1801 as a duodecimo volume and was priced at four shillings. It contained biographical accounts, some accompanied by illustrations, of nearly one hundred persons, beginning at the start of the alphabet with the eleventh-century French physician and philosopher Peter Abelard, remarkable for having been emasculated (by the relatives of his lover Héloise), and ending with the North American Quaker Jemima Wilkinson, who became a preacher and religious leader following her purported resurrection from the dead in 1776. In 1803, *Eccentric Biography* was joined by a companion, *Eccentric Biography: Or, Memoirs of Remarkable Female Characters, Ancient and Modern. Including Actresses, Adventurers, Authoresses, Fortune-tellers, Gipsies, Dwarfs, Swindlers, Vagrants, and Others Who Have Distinguished Themselves by Their Chastity, Dissipation, Intrepidity, Learning, Abstinence, Credulity, &c. &c. Alphabetically Arranged. Forming a Pleasing Mirror of Reflection to the Female Mind.*[46] Aimed at 'Ladies in particular', the second work contained only biographies of women. It ended, like the first, with Jemima Wilkinson, but opened with Alice, a native American slave who died in 1802 at the remarkable age of 116, then Joan of Arc.

These works were intended to be 'both interesting and useful'. A longstanding justification for publishing biography was that it could serve a useful moral purpose. In 1750, for example, Samuel Johnson wrote in his *Rambler* no. 60, 'On Biography', 'No species of writing seems more worthy of cultivation than biography, since none can be more delightful or more useful, none can more certainly enchain the heart by irresistible interest, or more widely diffuse instruction to every diversity of condition.'[47] In the preface to the second volume of *Eccentric Biography*, the editor expanded on the instructive potential of the genre, arguing that, as the book formed 'a pleasing mirror of reflection to the female mind', it would provide ladies with an opportunity of '*knowing themselves*'. Quite how 'ladies' were supposed to know themselves by reading about 'notorious Swindlers', 'Women of astonishing Weakness and Credulity', 'whimsical old Maids' and 'celebrated Courtezans' (as well as 'Ladies of wonderful fortune, enterprise and courage') was not spelled out. The majority of these 'eccentric' women cannot have been intended to function as role models, yet it was evidently felt that moral lessons could be drawn from contrasting the extremes by which certain remarkable women had deviated from the norms of womanly behaviour.

The following decades witnessed a proliferation of similar works: *The Eccentric Mirror: Reflecting a Faithful and Interesting Delineation of Male and Female Characters*, published in four volumes in 1806; the four-volume *Eccentric Magazine; Or Lives and Portraits of Remarkable Characters* (1812); *Kirby's Wonderful and Eccentric Museum; Or, Magazine of Remarkable Characters* (1803–20) in six volumes (see Figure 1.1); Thomas Tegg's single-volume *Eccentric Biography; Or Lives of Extraordinary Characters* (1826); Frederick Fairholt's *Remarkable and Eccentric Characters, with Numerous Illustrations* (1849); William Russell's *Eccentric Personages* (1864); John Timbs's *English Eccentrics and Eccentricities* (1866); and there were many others that appeared in this period.[48] Like their precursors, these works claimed to be both entertaining and instructive, frequently citing Alexander Pope's maxim, 'The proper study of mankind is man.'[49] Here

KIRBY'S

WONDERFUL

AND

ECCENTRIC MUSEUM;

OR,

MAGAZINE

OF

REMARKABLE CHARACTERS.

INCLUDING ALL THE

CURIOSITIES OF NATURE AND ART,

FROM THE REMOTEST PERIOD TO THE PRESENT TIME,

Drawn from every authentic Source.

———◆———

ILLUSTRATED WITH

ONE HUNDRED AND TWENTY-FOUR ENGRAVINGS.

CHIEFLY TAKEN FROM RARE AND CURIOUS PRINTS

OR ORIGINAL DRAWINGS.

———◆———

SIX VOLUMES.

VOL. IV.

LONDON:

R. S. KIRBY, LONDON HOUSE YARD, ST. PAUL'S.

——

1820.

Figure 1.1. Kirby boasted that in his *Eccentric Museum*, readers could discover 'All the Curiosities of Nature and Art from the Remotest Period to the Present Time'. Mixed in with biographical sketches, he included accounts of ghosts, mermaids, multiple births, remarkable frosts, dreams, explosions and a host of other remarkable phenomena. Reproduced by kind permission of the Syndics of Cambridge University Library.

I trace the development of eccentric biography, outlining some of the defining features of the genre. I argue that, more than merely being an indicator of the prevalence of a discourse of eccentricity in the first half of the nineteenth century, eccentric biography served, first, to propagate this discourse amongst ever diversifying readerships and, second, to popularize new ways of organizing and presenting biographical material about remarkable persons, thus shaping how eccentricity was subsequently understood and represented.

Collective biography had its roots in antiquity, but blossomed in the age of Enlightenment, a period marked for its production of encyclopaedias and dictionaries, including the first dictionary of national biography, the six-volume *Biographica Britannica* of 1744.[50] Particularly influential was James Granger's *Biographical History of England, from Egbert the Great to the Revolution* (1769), an exhaustive compilation of short biographies arranged firstly by monarchic reign and then hierarchically by class.[51] There were twelve classes, ranging from 'Kings, Queens, Princes, Princesses, &c. of the Royal Family' in Class I, through lesser nobility and commoners of distinguished employment, down to 'Physicians, poets, and other ingenious persons' in Class IX, 'Painters, Artificers, Mechanics' in Class X, and women in Class XI. The final class dealt with 'Persons of both Sexes, chiefly of the lowest Order of the People, remarkable from only one Circumstance in their Lives; namely, such as lived to a great Age, deformed Persons, Convicts, &c.' The key to the popularity of the work was that it catalogued known portraits of each of the biographical subjects. The volumes were not illustrated, but many owners chose to extra-illustrate their personal copies, interleaving prints purchased singly or taken from other books. Granger's *Biographical History* was implicated in the print-collecting craze which swept Britain in the late eighteenth century, and the practice of binding extraneous but thematically related material into books became known as 'grangerizing'.[52] As the following extract, from a poem of 1813, illustrates, the effects of grangerism continued to be felt well into the nineteenth century:

> *Granger* – whose biographic page,
> Hath prov'd for years so much the rage,
> That scarce one book its portrait graces,
> Torn out alas! each author's face is.[53]

Many of the prints catalogued by Granger were extremely rare, and it was not long before ready illustrated versions of the *Biographical History* began to appear. One such was a *Catalogue of Richardson's Collection of English Portraits, Engraved from Rare Prints, or Original Pictures: – viz Princes and Princesses, Secretaries of State ... as Described in Granger's Biographical History of England*, published by W. Baynes in the 1790s.[54] Another was *Portraits, Memoirs, and Characters of Remarkable Persons from the Reign of Edward the Third to the Revo-*

lution, compiled and published by the print-seller James Caulfield in 1794, and republished as a second edition by R. S. Kirby in 1813.[55] Published galleries of portraits – books such as the *Biographical Magazine*, 'containing Portraits and Characters of Eminent and Ingenious Persons of every Age and Nation' (1794), the *Biographical Mirrour* (1795) and the *British Cabinet*, 'containing Portraits of illustrious personages, engraved from original pictures: with biographical memoirs' – were common by the 1790s.[56] Caulfield's *Portraits* was particularly significant for the history of eccentricity because it served as a link between the older genre of collected portraits and the emergent genre of eccentric biography. *Portraits* was intended as a completion of Granger's twelfth class – of the 'lowest Order of the People' – this being the class with which Granger had experienced most difficulty in locating portraits. In keeping with the fashion for collecting portraits of criminals and vagabonds, Caulfield had in 1795 published *Black-guardiana*, a 'Dictionary of Rogues, Bawds, Pimps, Whores ... illustrated with eighteen Portraits of the most remarkable Professors in every Species of Villany', and he drew upon this work in completing *Portraits*.[57] *Portraits* did not announce itself to be concerned specifically with 'eccentric' subjects; furthermore, it addressed different audiences to those addressed by later volumes of eccentric biography. Nevertheless it was cited explicitly as being 'the cause' of at least one collection of eccentric biography, and the biographies and engravings it contained were drawn upon in countless subsequent works.[58]

Literary genres have traditionally been defined in terms of the properties of texts rather the properties of books. However, the physical composition of a book – its format, typesetting, illustrations and so on – can also be important in determining which genre or genres it belongs to.[59] Granger's *Biographical History* and Caulfield's *Portraits* were both expensive, luxurious works aimed at wealthy readers and book collectors. *Portraits* was originally sold in 1794 for a hefty 2*l*.10*s*. for two volumes; by 1813 the work had become so rare that copies were allegedly being sold for the enormous sum of seven guineas, prompting Caulfield to produce a second edition.[60] The rarity and consequent expense of this sought-after work prompted enterprising publishers to produce more accessible alternatives. These successors began to adopt the epithet 'eccentric' in their titles, and tended to aim their books at a less affluent, middle-class readership. The *Eccentric Mirror* (1806), for instance, though costing a total of 22*s*. for a complete four-volume set, was published in parts of approximately 35 pages, making it accessible to readers with a modest income; the *Eccentric magazine*, produced with the cooperation of Caulfield, was still relatively expensive in 1812 at 16*s*.6*d*., but reassured potential purchasers that this represented value for money when compared with the cost of acquiring a copy of *Portraits*.[61] In later decades, eccentric biography became cheaper and cheaper: Thomas Tegg's unillustrated *Eccentric Biography* was sold in 1826 for 4*s*.6*d*. and Fairholt's *Remarkable and Eccentric Characters* cost just 2s.6d. in 1849. There were exceptions to the trend, but while both Russell's *Eccentric Personages* (1864) and Timbs's *English Eccentrics and Eccentricities* (1866), for example, were initially sold for 21s., cheaper editions were published soon afterwards.

The rise of eccentric biography was part and parcel of a much broader set of changes in how books were produced and made available to the various constituencies of the growing reading public. In *The Reading Nation in the Romantic Period*, William St Clair argues convincingly that book production and reading increased rapidly in the period from around 1790 to 1830 largely as a result of changes to copyright legislation.[62] His argument represents a departure from previous histories, which have identified primary education and technological developments in printing as the main drivers of the increase. Although perpetual copyright was officially brought to an end in 1710 with the so-called Act of Queen Anne, it was only after a House of Lords judicial decision in 1774 that the Act was enforced in practice. Before 1774, the first person to print (rather than to author) a text retained intellectual property in perpetuity, so it could never legally be reprinted by anybody else. Under this monopoly system, prices of books were deliberately kept high; thus most books remained out of the reach of all but the wealthiest readers. From 1774 until a further Act in 1808, copyright was limited to fourteen years, plus another fourteen if the author was still alive at the end of the period. The result, St Clair argues, was an explosion in the production of books, many of them reprints of older, now out-of-copyright titles. As a new, competitive marketplace emerged, books became smaller and cheaper. The pressure to offer lower prices fuelled investment in reducing manufacturing costs through the invention and harnessing of new technologies: stereotyping, machine-made paper, ready-made bindings and steel engraving, for example.[63] Even under these circumstances, the cheapest out-of-copyright books were too expensive for the majority of the population to purchase outright. But the book industry responded by selling books in parts, or 'numbers', which were distributed around the country by travelling salesmen.[64] For the first time, a wide range of recently authored literary works was made available to readers from the middling and even the lower ranks of society, whose reading, up until that time, would have mostly been restricted to almanacs, ballads and the Bible. The trend of lowering prices and the innovation of selling books in numbers are both reflected in the pricing of eccentric biography.

As books began to be printed in much larger print runs, it became economically viable to sell illustrated books at moderate prices. Moreover, from the 1830s, woodcut and steel engraving enabled tens of thousands of impressions to be made from a single plate; under the older system of copper engraving, the plates usually wore out after a few thousand imprints.[65] Illustrated books became increasingly sought after by middle- and working-class readers. Illustrated biography was perfectly adapted to this market, and rival editors stressed the number, quality, rarity and originality of the pictures contained within their works. Generally speaking these were copied, with variable degrees of skill and attention to detail, from existing portraits, especially from Caulfield's *Portraits*. Alternatively, old woodcuts from previous publications were simply reused, thus reducing production costs even further.

The changes to the intellectual property regime in the 1790s affected not only the pricing of books, but also the types of books that were produced. For example, perpetual copyright had severely restricted possibilities for publishing adaptations or collections of extracts from works by different authors. Once perpetual copyright was lifted, however, there was a boom in anthologies, abridgements and simplified and censored versions of out-of-copyright works.[66] These played an important role in making texts available to wider readerships, including poorer and less well educated readers, and children. A further development was the rise of the 'annual'. During the eighteenth century a fashionable activity amongst ladies was the making of 'commonplace books': decorative albums containing cut-and-pasted printed text and images, *sententiae* and handwritten passages of prose and poetry written down from memory or copied from books. The practice, which formed part of an upper-class culture of 'feminine accomplishments', persisted into the nineteenth century. However, from around 1820 there was a craze for printed annuals or 'gift books', some interleaved with blank pages. With titles such as *The Keepsake*, *Forget-Me-Not*, *Friendship's Offering* and the *Drawing Room Scrap Book*, they were particularly popular as Christmas presents; some titles sold as many as 15,000 copies a year.[67]

The development of new 'miscellaneous' kinds of book was not restricted to the realm of imaginative literature. Eccentric biography similarly brought together chunks of diverse material from a variety of sources. An important feature of the genre was that it allowed for each volume or pamphlet to be composed of short, well defined sections of text, each relating to a different biographical subject. Eccentric biography was thus particularly suited to readers with limited time or literacy skills. Volumes could be read over many short sessions without loss of continuity; alternatively they could be used as 'dipping-in' books. Eccentric biography thus intersected with a number of other non-fictional genres aimed at readers from the lower to middling orders of society: miscellanies such as Charles Hulbert's *Breakfast of Scraps* (date unknown) and Burkett's and Plumpton's *Cabinet of Curiosities* (c. 1810), encyclopaedic compilations such as Hulbert's *Select Museum of the World* (1822), and cheap periodicals, such as the two-penny *Mirror of Literature* (1822–49).[68]

The *Mirror of Literature* was edited between 1827 and 1838 by John Timbs (1801–75), who drew heavily on his earlier writings in compiling *English Eccentrics and Eccentricities* in the 1860s.[69] Up until the early decades of the

nineteenth century, writers lacking an independent income had generally relied
on patrons to support them financially. With the development of mass reading audiences, however, writing emerged as a commercially viable activity, and
Timbs is an example of an early professional 'journalist' who (with considerable difficulty) supported himself by writing for magazines and newspapers, and
by producing volumes of edited miscellany aimed at mass middle- and working-class readerships. Timbs was notorious amongst his contemporaries for
his use of 'scissors-and-paste' editorial techniques, whereby old print material
was digested and recombined in order to produce 'new' books – the publisher
Henry Vizetelly described Timbs as having 'spent the best part of a busy life,
scissors in hand, making "snippets".'[70] From its inception, eccentric biography
made routine use of scissors and paste. While editors stressed the novelty of the
material they presented to the public, they also placed emphasis on the reliability of their sources, thus implicitly acknowledging their roles as compilers rather
than original authors. The preface to the *Eccentric Mirror*, for example, assured
readers that the editor had obtained 'a large portion of curious original matter', whilst also boasting that 'many valuable and expensive foreign publications'
had been consulted in its preparation. *Kirby's Wonderful and Eccentric Museum*
likewise boasted 'a very competent share of Original and Eccentric Biography'
alongside 'Translations from the most scarce and valuable Originals in the modern languages'. Fairholt acknowledged having drawn on the *Eccentric Magazine*,
Kirby's and others, but claimed that 'some information has in all instances been
added, which will not be met with in previous compilations'. Timbs, by contrast, preferred to copy his authorities word for word. By way of justification
he argued that 're-writing … rarely adds to the veracity of story-telling, but, on
the other hand, often gives a colour to the incidents which the original narrator
never intended to convey'.[71]

In a period when writing was becoming increasingly commercialized, eccentric biography lay at the heart of debates about literary expertise and plagiarism,
fuelling growing concerns about the quality, in particular the moral quality, of
the literature that was being consumed by the new mass reading publics. Critics
of eccentric biography argued that persons responsible for such works were compilers rather than 'authors', and maintained that their works were unworthy of the
name of 'literature'.[72] Henry Fothergill Chorley, reviewing Fairholt's *Remarkable
and Eccentric Characters* for the *Athenaeum*, acknowledged the inherent fascination of eccentric biography, but asserted that 'By no one … could the subject have
been handled with much less research than by Mr. Fairholt'. The book, he complained, was 'a ten-times told tale'; the stories were so over-familiar he could have
repeated them by heart.[73] Such denigrations tended to be all the harsher when
'compilers' made only feeble attempts to organize their material. 'The efforts at
classification are neither systematic nor successful', John Jeaffreson complained

of Timbs's *English Eccentrics and Eccentricities*. 'Thus, under the head of "Wealth and Fashion" we have one hundred and thirty pages of familiar, though occasionally mistold and altogether disconnected, stories about the Beckfords, Beau Brummell, John Ward the Hackney Miser ... and half-a-hundred dissimilar notabilities. Under the comprehensive title, "Delusions, Impostures and Fanatic Missions", just as many pages are given to anecdotes about a number of men, women, and things that have no natural connexion.'[74]

The insults multiplied. 'He does not use his collected materials to illustrate a subject or a series of subjects, but turns out his rag-bag in a heap, and leaves his readers to choose', wrote John Doran of Timbs's *Things Not Generally Known*.[75] 'It is an example of "paste-and-scissors" labour', wrote Frederick Stephens of Timbs's *Anecdote Biography*; 'the masses of notes and comments are pitched together, without the slightest effort at selection or even examination into the authenticity of the matter reproduced from all sorts of sources'.[76] Perhaps the most damning review of all was of *English Eccentrics and Eccentricities*:

> Unfortunately for Mr Timbs, we cannot credit him with the possession of any one of the qualities in the absence of which his low art has no claim to respect. He gathers stories with his right hand and his left, cutting them from newspapers and spurious biographies, from discredited memoirs and worthless chapbooks; and when he has filled his basket full of miscellaneous waifs and strays, he throws the undigested mass of rubbish at the feet of the public, just as a dustman discharges the contents of his cart upon the contractor's heap, and then moves away, without caring to analyze his contribution to the unsightly mound ... Upon the whole, a worse book of its bad kind has not appeared for some long time.[77]

Poor Timbs! Such crushing criticism was to become matter of course throughout his long and extraordinarily productive career as a journalist and compiler.

Cut-and-paste techniques have been used in the [re]production of eccentric biography to this day. The most famous twentieth-century volume of eccentric biography is Edith Sitwell's *The English Eccentrics*, which has been reprinted several times since it was first published in 1933.[78] Since the 1980s especially, countless further volumes have been published: The Reader's Digest's *Great British Eccentrics* (1982); John Keay's *Eccentric Travellers* (1982); Peter Bushell's *Great Eccentrics* (1984); Dutton and Nown's *Oddballs! Astonishing Tales of the Great Eccentrics* (1984); Margaret Nicholas's *The Worlds Greatest Cranks and Crackpots* (1984); John Michell's *Eccentric Lives and Peculiar Notions* (1989); *Timpson's English Eccentrics* (1991); Clifford Pickover's *Strange Brains and Genius: The Secret Lives of Eccentric Scientists and Madmen* (1999); Karl Shaw's *Mammoth Book of Oddballs and Eccentrics* (2000); John Joliffe's *Eccentrics* (2001); and so on, and so on.[79] The Beckfords, the misers, Beau Brummell and a host of persons to be introduced in the following sections are all still there, alongside more recent additions. Nineteenth-century eccentric biography established a canon of eccentric individuals whose anecdotal life-stories became increasingly well known as the origins of those stories became increasingly obscure.

Moreover, as a result of cut-and-paste literary techniques, eccentricity came to be associated with a particular literary style: a style defined by short, circumscribed narratives, jumbled cumulatively together. *Anecdotes* became the building blocks of a discourse of eccentricity which proceeded according to a logic of addition: such and such an eccentric character was very wealthy, and very reclusive, and very miserly, and lived to a great age, and once dressed up as a beggar, and once hid under a table and barked like a dog, and once met Samuel Johnson, &c., &c., &c.

The rise of eccentric biography can be understood as one response to the problem of imposing order upon those individuals who did not fit comfortably within what George Cruikshank called, in 1840, 'The British Beehive' (see Figure 1.2). In Cruikshank's cartoon, representatives of the various trades are graphically allocated within their respective compartments in a vast architectural structure representing the British industrial economy. The governing institutions of the state, and the arts and professions are supported by commerce and industry, which in turn rest upon the efforts of the labouring poor. Within this structure, there is no room for hermits, misers, vagrants, gypsies, deformed persons, criminals, prostitutes, impostors, the old, the mad, or any persons directing their efforts towards impractical ends. While comic compilations such as Albert Smith's *The Natural History of Stuck-up People* and Angus Reach's *The Natural History of Bores*, both of 1847, satirized this persistent obsession with classifying people, books such as the antiquarian John Smith's *Vagabondiana: Or, Anecdotes of Mendicant Wanderers Through the Streets of London* of 1817, or the many Cries of London (anthologies of street-vendors' calls) published during this same period, domesticated vagrant activity.[80] Within this context, discourses of eccentricity served a complex social function, on the one hand defining the limits of socially acceptable behaviour, on the other containing transgressive behaviours by subjecting them to rational taxonomic order.

Challenging Boundaries

The problem of ordering had challenged eccentric biographers since the inception of the genre and seems never to have been satisfactorily resolved. Cundee ordered his volumes of 1801 and 1803 alphabetically without any explicit justification. His successors tended instead to adopt a logic of no logic, maximizing the differences of character between juxtaposed individuals, thus enhancing the miscellaneity of the compilation. In 1849, Frederick Fairholt made an explicit case for this tried-and-tested arrangement. It might be possible, he observed, to arrange such a work in sections illustrative of specific historical periods or coun-

Figure 1.2. George Cruikshank designed 'The British Beehive' in 1840, revising and reprinting it in 1867. The image represents British industrial society as a highly ordered heirarchy, in which each individual's contribution is essential to the well-being of the whole. © V&A Images/Victoria and Albert Museum.

tries, but this would require 'heaviness of detail'. Alternatively, persons might be organized into groups representative of 'particular kinds of eccentricity', yet such an arrangement would bring with it the disadvantage of monotony. The benefit of juxtaposing dissimilar persons was that it 'brings each character out more ... remembering that eccentricity itself is in the last degree miscellaneous and irresponsible and that any attempt to methodize or catalogue its specialities, must, to a certain extent, be a failure, not merely on the score of philosophy, but of amusement'.[81] As Fairholt recognized, ideas about order and classification were fundamentally important to how eccentricity was understood in this period.

In 1869, Timbs published a companion to *English Eccentrics and Eccentricities* entitled *Eccentricities of the Animal Creation*. In this book Timbs arranged animals under the section headings 'The rhinoceros in England', 'Stories of mermaids', 'Is the unicorn fabulous?', 'The mole at home', 'The great ant-bear', 'Curiosities of bats', 'The hedgehog', 'Eccentricities of penguins', 'Pelicans and cormorants', 'Talking birds, etc.', 'Owls', 'Weather-wise animals', 'The tree-climbing crab', 'Musical lizards', 'Chameleons and their changes', 'Running toads', 'Frog and toad concerts', 'Song of the cicada', 'Stories about the barnacle goose', 'Leaves about bookworms' and 'Boring marine animals, and human engineers'.[82] One of the aims of natural history in the nineteenth century was to appropriate diverse kinds of natural object within taxonomic frameworks; however, natural history was also associated with a longstanding tradition of collecting and investigating anomalous natural objects or 'monsters'.[83] Timbs's eccentric animals resisted classification: the mermaid was half human half fish; the tree-climbing crab was a crab, but it climbed trees; the bookworm was a worm, but it lived in books; and so on. These animals were labelled 'eccentric' because they appeared to transgress the classificatory boundaries which were meant to order animals into different kinds.

As the grand finale to his *Wonderful and Eccentric Museum*, Kirby exhibited the hermaphrodite Mademoiselle Lefort, 'the most singular phenomenon that has occurred in our day'.[84] Kirby found himself 'at a loss whether to describe this human being by the term masculine or feminine'. Calling herself 'Mademoiselle' but usually electing to wear men's clothing (cf. Figure 1.3, in which she is wearing a dress), Lefort had made a small fortune exhibiting herself as a public spectacle. She also allowed herself to be examined by members of the medical profession, to whom, Kirby knowingly assured his readers, she proved an 'inexhaustible source of professional inquiry'. Kirby described Lefort in language appropriate partly to a natural-historical specimen, partly to an artistic creation: 'The hands, arms, feet, and bust', he wrote, 'possess perfect feminine beauty, likewise the upper part of the face; the lower part is also beautiful, but possessing all the masculine accompaniments, as mustachoes and whiskers, of very strong black hair, with which the chin is also thinly sprinkled.' The caption to Kirby's illustration reads 'Mademoiselle Lefort, Exhibited in Spring Gardens 1818'. In his abandonment of the generic conventions of biographical representation, Kirby objectified the subject of his 'biography', such that rather than being an active, complete human being she became a passive amalgam of body parts, an 'exhibited' object or specimen to be viewed with a quasi-scientific curiosity.

MADEMOISELLE LEFORT.
Exhibited in Spring Gardens, 1818.

Figure 1.3. Lefort's fashionable empire-line dress shows off her feminine bust, whilst her flowing skirts leave other anatomical details to the imagination. From *Kirby's Wonderful and Eccentric Museum* (1820), vol. 6. Reproduced by kind permission of the Syndics of Cambridge University Library.

Mathew Robinson,

LORD ROKEBY,

Aged 88.

Pub.d June 28.th 1805 by R.S.Kirby 11 London House Yard St Pauls.

Figure 1.4. Rokeby's beard was not his only eccentricity. His clothes were plain, even mean, and in an age of luxury he was abstemious in eating and drinking. Eschewing the fashionable landscape gardening of the eighteenth century, he left his estate to vegetate in wild luxuriance. From *Kirby's Wonderful and Eccentric Museum* (1820), vol. 3, p. 379. Reproduced by kind permission of the Syndics of Cambridge University Library.

The popularity of 'freak shows' in the nineteenth century is well known: giants, dwarves, deformed persons and persons with extraordinary physical capabilities were commonly exhibited in shows, circuses and exhibitions, and constituted a significant portion of the shows of London.[85] Many of these same people also featured in the pages of eccentric biography. Jeffrey Hudson the celebrated dwarf, Daniel Lambert the fat man, Thomas Topham the strong man, Matthew Buchinger, who was born without hands, feet or thighs, Patrick O'Brien the Irish giant – these and many more were exhibited on paper as *lusus naturae*.[86] Volumes of eccentric biography were printed equivalents of live displays of human wonders: paper menageries or 'museums', as Kirby called his collection. Like Timbs's eccentric animals, these 'eccentric' persons physically challenged fundamental boundaries: between male and female, adult and infant, human and non-human.

While some human beings were labelled eccentric on the basis of physical characteristics, in other cases it was on the basis of their habits. Examples from eccentric biography could be multiplied almost indefinitely; I shall give just a handful here by way of illustration. Individuals who appeared to belong to another age, generally an ill-defined historical past, featured prominently in eccentric biography. A popular example, appearing in at least five of the collections previously mentioned, was Lord Rokeby (1712–1800). Rokeby was labelled 'eccentric' on account of his beard 'whose length, for it reached nearly to his waist, proclaimed it of no recent date' (see Kirby's illustration in Figure 1.4).[87] Kirby explained:

> Beards were once considered as marks of respectability, particularly among the ancients. With regard to this article, however, opinion is now reversed, and it is, at least regarded as an indubitable token of eccentricity. Why it was adopted by his Lordship is not known; reasons for such conduct are not easily discovered; it bids defiance to conjectures, and baffles all sagacity.[88]

Rokeby's beard was eccentric because it was dated; its existence challenged the grooming conventions of the age. However, the beard itself is an indicator of time, and the extraordinary length of Rokeby's beard further signified that Rokeby had resisted change over many years. Rokeby's beard was a 'token' of eccentricity because it doubly signified that its owner lived in one age, yet belonged to another.

By the time Timbs came to write *English Eccentrics and Eccentricities* in 1866, opinion on beards had reversed once more, but Timbs identified other modern characters whose appearances evoked a past way of life. Take, for example, Mrs Lewson of Clerkenwell, commonly called 'Lady Lewson' on account of her 'eccentric' manner of dress. Born in 1700, Lewson was so fond of the fashions of her youth that she continued to wear them long after they had ceased to prevail.

More
Fashionista

She generally wore a silk dress, with a long train, a deep flounce all round, and a very long waist; her gown was very tightly laced up to her neck, round which was a ruff or frill; the sleeves came down below the elbows, and to each of them four or five large cuffs were attached; a large bonnet, quite flat, high-heeled shoes, a large black silk cloak trimmed with lace, and a gold-headed cane, completed her every-day costume for eighty years.[89]

The key features of Lady Lewson's dress, described here in the most intricate detail, were, first, that it was of an antique style and, second, that, like her 'very methodical' habits, it had never changed. Lewson was perceived to be eccentric because her dress and her habits did not keep pace with the times. While the rest of the world moved on, she continued to live in a previous age. Her inherent resistance to the passage of time was confirmed by the fact that she died only in 1816, aged 116.

Two further stock kinds of 'eccentric' character, sometimes overlapping, were misers and hermits. Included in the class of misers were W. Fuller, Esq., 'a pecunious character'; Daniel Dancer, 'an extraordinary miser'; the 'celebrated Miser', John Elwes, Esq., also known as 'the British miser – *par excellence*'; William Jennings; M. Ostervald, the French banker who died of want; John Ward, the Hackney Miser; James Taylor, the Southwark Miser; and Mr Thomas Cooke, the celebrated Miser of Pentonville. Celebrated hermits included Henry Welby, an 'extraordinary character', who lived forty-four years the life of a hermit in the city of London; Sarah Bishop, 'the female hermit'; Roger Crabb, 'a singular hermit'; John Bigg, the celebrated Denton Hermit; and a great many professional ornamental hermits who were employed to decorate the estates of wealthy landowners.[90] Misers challenged notions of rich and poor: frequently extremely wealthy, they chose to live impoverished lives, denying themselves basic necessities as well as the luxuries they could easily afford. Hermits denied themselves the pleasures of human company; alone and isolated they existed, like beggars, vagabonds and criminals, at the margins of social life.

Eccentricity also arose in discussions of the boundary dividing sanity from madness. In July 1787 the word 'eccentricity' appeared in *The Times* in a report of the trial of a London apothecary John Elliot, accused of 'unlawfully, wickedly, and maliciously shooting at' a young woman, Mary Boydell.[91] Elliot's defence was not that he had not committed the crime; rather it was that he was 'not, by the law of England, accountable for his actions'. One of the witnesses in his defence, an apothecary who had previously been in partnership with him, testified that, though the defendant had always treated his patients with propriety and had poisoned none of them, 'there was always an eccentricity about him'. Lately he had 'been wandering about from place to place, and [had] had no connection with any body'. A second witness, Dr Simmons, stated that he believed the defendant to have been 'disordered in his mind' for some months. Simmons explained that the defendant 'lived in a room where no body visited him, and made a great number of experiments'. He then produced a paper, written by the defendant, endeavouring to show that the sun was not a body of fire, but might be a commodious habitation. The paper was placed before the court as evidence of the defendant's insanity but, unfortunately for the defendant, the Recorder thought the paper 'ingenious'. The trial subsequently segued into a discussion of Buffon's theory of the formation of the Earth.

In 1787 a vaguely defined eccentricity of manner could be put forward at court as evidence of a person's insanity. In later years, however, eccentricity and insanity came to be opposed to one another in matters of law. In April 1826, *The Times* reported the judgement of Dew vs Clark and Clark, an 'important, difficult, and novel case' concerning the contested will of the surgeon and medical electrician Elias Stott.[92] The suit was commenced in April 1822 by Stott's daughter, Charlotte Dew, who alleged that her father had been insane when he had drawn up his will leaving his considerable fortune to his nephews Thomas and Valentine Clark, rather than to herself. Clark's and Clark's defence of the will was that 'although the testator might have shown an eccentricity of mind, proceeding from an original defect of education, a violent temper, and the effects arising from an overweening partiality to Methodism, or Calvinistic doctrines of religion', not to mention his 'absurd' notions respecting electricity, he was, 'on general topics sound; and the law knew no such thing as partial insanity'. The case was made difficult by Stott's having treated his daughter with brutality during his lifetime, but the question for the court was simply whether the testator was of sound or unsound mind. 'Eccentricity, severity, and violence were not enough to establish insanity', explained *The Times* court reporter, and the first task of the court was thus 'to define where eccentricity ended, and derangement commenced'.

Distinguishing between 'mere eccentricity' and 'unsoundness of mind, or insanity' became a matter for medical and legal experts.[93] In Dew vs Clark and Clark, recent medical treatises were consulted, as were the writings of John Locke. In other cases medical men were summoned to give their opinions on the case at hand. 'The medical attendant ... thought that the defendant's state of mind bordered closely upon madness', explained a *Times* reporter covering the lunacy trial of an unnamed Spaniard, 'but said he could not take upon himself to call it lunacy, but rather eccentricity'.[94] 'I decidedly think that he is of perfectly sane mind', concluded a physician giving evidence in the case of the Reverend Edward Frank, an alleged lunatic, adding on a cautionary note, 'His eccentricity is such that, together with dimness of sight, it might lead a person less experienced in examining the state of the human mind to conclude that he is insane.'[95] In the early nineteenth century eccentricity occupied a contested space at the juncture of madness and sanity, functioning as a foil against which both madness and 'normality' could be defined.[96]

Eccentricity, Gender and 'Separate Spheres'

The contested relationship between eccentricity and 'normality' can be further illuminated through an analysis of eccentricity and gender. Since the 1980s scholars have debated gender roles in the eighteenth and early nineteenth centuries in terms of 'separate spheres'. Particularly influential has been Davidoff's and Hall's argument that, with the emergence of a defined middle class, women were excluded from the public sphere and confined to the private sphere of the home.[97] This argument has been criticized from several angles. Amanda Vickery, for example, has argued that a separation of spheres occurred much earlier than Davidoff and Hall allow.[98] Lawrence E. Klein and others have challenged the public/private dichotomy at an analytical level by pointing to the wide range of meanings held by the terms 'public' and 'private' in the period.[99] Attention has furthermore been drawn to the reliance of separate-spheres models on prescriptive literature, which, critics insist, cannot be assumed to reflect how men and women really behaved. As a result of this constructive debate, it is now generally agreed that, in practice, there was more overlap between men's and women's roles than acknowledged in early separate-spheres models.[100] Nevertheless, in nineteenth-century novels and eccentric biography – both of which can be understood, in a sense, as forms of prescriptive literature – men and women were labelled 'eccentric' if they engaged in activities believed to be inappropriate to their sex, that is, to use the astronomical metaphor, if they were seen to stray beyond their proper spheres. I shall here discuss just a couple of examples.

Female eccentricity was often explicitly equated with masculinity. A large part of eccentric biography, for example, was devoted to women who disguised themselves as men in order to pursue particular careers. 'Eccentric' women who adopted the male attire included Hannah Snell (1723–92), the female soldier, who rejected marriage proposals from suitors of both sexes; Charlotte Charke (d. 1760), an actress who tried to pass herself off upon the world as a man; Mary Frith, thief, pickpocket, fortune-teller and male impersonator, also known as Moll Cutpurse; Mary Anne Talbot, the sailor, also known as John Taylor; Mrs Christian Davies, the female soldier, commonly called Mother Ross; Renée Bordereau, or, Langevin, the military heroine of Vendée; Phoebe Bown, the Irish strongwoman; Hester Hammerton, the female sexton; and Mary East, also known as James How, the 'female husband'.[101] These women challenged gender norms to the extreme, choosing to wear the clothing of the opposite sex, and successfully fulfilling the social, professional and sexual roles allotted to them on the basis of their assumed masculine identities.

Other women who featured in eccentric biography exhibited an outwardly feminine demeanour, but were described as possessing 'eccentric' masculine character traits, such as a 'masculine understanding' (attributed to the unmarried mother Polly Baker), or a 'daring and masculine spirit' (attributed to Fulvia, wife of Mark Anthony). One somewhat overdetermined example is that of Christina, Queen of Sweden.[102] Christina's birth in 1626 was attended with confusion: the royal astrologers had predicted that she would be a boy, and some 'false appearances' initially convinced the royal household that the prediction had been satisfied. When the truth became apparent the king, rather than being disappointed, rejoiced, announcing, 'I trust this girl will be as good as a boy; she must certainly be clever, for she has deceived us already.' From her infancy, Christina demonstrated an 'unconquerable aversion ... for the yoke of marriage'. She developed 'an invincible antipathy for the employments and conversation of women', and displayed 'the natural aukwardness of a man' when confronted with women's work. Rejecting conventional women's leisure activities, she instead indulged herself in 'violent exercises' and 'abstracted speculations', amusing herself especially with language, legislature and the sciences.

Numerous further examples of 'eccentric', 'masculine' female characters are to be found in novels. In Maria Edgeworth's *Belinda* (1801), for example, Mr Vincent describes Mrs Freke as 'a dashing, free-spoken, free-hearted sort of eccentric person, who would make a staunch friend, and a jolly companion'. 'As a mistress or a wife', he adds, 'no man of any taste could think of her.'[103] In Charlotte Brontë's *Shirley*, Shirley Keeldar interrogates her tenant, Gérard Moore, as to why he finds Caroline Helstone 'peculiar':

> '... knowing her [Shirley says], I assert that she is neither eccentric nor difficult of
> control: is she?'
> 'That depends –'
> 'However, there is nothing masculine about her? ... Caroline is neither masculine, nor
> of what they call the spirited order of women.'[104]

In Charles Dickens's *David Copperfield* (1850), David's aunt is portrayed as masculine and eccentric on account of her predilection for vigorous physical exercise and her intellectual abilities:

> I know that my aunt distressed Dora's aunts very much, by utterly setting at naught
> the dignity of fly-conveyance, and walking out to Putney at extraordinary times,
> as shortly after breakfast or just before tea; likewise by wearing her bonnet in any
> manner that happened to be comfortable to her head, without at all deferring to the
> prejudices of civilisation on that subject. But Dora's aunts soon agreed to regard my
> aunt as an eccentric and somewhat masculine lady, with a strong understanding.[105]

In Charlotte Mary Yonge's *The Daisy Chain; Or, Aspirations* (1856), Miss Winter, the governess, worries that young Ethel is 'walking continually in all weathers', and is 'at every spare moment busy with Latin and Greek'. Ethel is deficient in feminine accomplishments – her dress, hair, needlework and handwriting are shameful – and Miss Winter warns that she will 'grow up odd, eccentric, and blue' if she is not prevented from overexerting herself.[106]

Gender boundaries were central to conceptions of male, as well as female, eccentricity. Eccentric biography reported cases of female impersonation, such as those of Elizabeth Russell, who was 'always known under the guise or habit of a woman' and yet was at death proved to be a man.[107] Most notorious was the case of Mademoiselle la Chevalière D'Eon de Beaumont (1728–1810) (see Figure 1.5). D'Eon's sex was unknown during his lifetime, despite a public trial to determine it, but he conducted diplomatic negotiations as both a woman and a man, and exhibited his fencing abilities at the Rotunda before the Prince of Wales whilst wearing a dress.[108] Avoiding the issue of possible moral depravity, commentators framed their criticisms in terms of wasted opportunity: 'On a survey of all the circumstances of the life of this truly extraordinary character, it is impossible to forbear lamenting the extreme perversity of the human mind, which sometimes impels an individual, for the sake of gratifying a foolish whim, to sacrifice all those advantages which, with the exercise of ordinary prudence, would lead to fortune and honourable distinction.'[109] In renouncing his sex, D'Eon also renounced all those advantages afforded him as a man.

Generally speaking, 'eccentric' men were not described as being 'feminine'; rather they were portrayed as not being masculine enough. Success within the world of work was central to nineteenth-century definitions of normative bourgeois manliness,[110] and men who recoiled from opportunities to shine in public life were labelled 'eccentric' on account of their refusal to participate in traditionally masculine spheres of activity. In 1823, for example, *The Times* reported the evangelical minister Thomas Chalmers's decision to give up his ministry of St John's in Glasgow in favour of a professorship of Moral Philosophy at St Andrews.[111] St John's was an overpopulated, impoverished parish and Chalmers achieved considerable public acclaim for revitalizing the parish church and strengthening its role in catering to the physical and pastoral needs of the parish poor. Upon his decision to move to St Andrews, *The Times* lamented 'the unaccountable eccentricity of views which could induce him to withdraw from a scene where his great zeal and splendid talents found so ample a field of useful exertion ... and retire to a place where all the influence of his name will not prevent his powers from being lost to the world'. In choosing to pursue abstract philosophical researches, Chalmers was seen to be jeopardizing his 'well-earned fame', and turning his back on his manly obligation to be socially 'useful'; he was deemed 'eccentric' because he was seen to be wilfully redirecting the course of his life away from traditionally masculine goals.

In the examples given above, the activities of 'eccentric' men are subjected to straightforward moral disapproval. However, eccentricity also featured in works more critical of stereotypical notions of manliness. In Maria Edgeworth's *Patron-*

Chevalier D'Eon de Beaumont,

Figure 1.5. After a trial in 1777 to determine his sex, d'Eon always wore female attire. Upon his death he was medically examined and pronounced to be a man. From *Kirby's Wonderful and Eccentric Museum* (1820), vol. 4, frontispiece. Reproduced by kind permission of the Syndics of Cambridge University Library.

age (1814), for example, Lord Oldborough questions Mr Percy about his family, and is surprised to discover that Percy, though apparently a man of talent, lives in virtual retirement. Despite Percy's insistence that he is fond of domestic life, Oldborough assumes that he would be glad to shine in public given the opportunity.

> Upon this supposition, his lordship dextrously pointed out ways by which he might distinguish himself; threw out assurances of his own good wishes, compliments to his talents, and, in short, sounded his heart, still expecting to find corruption or ambition at the bottom. – But none was to be found – Lord Oldborough was convinced of it – and surprised.
>   – Perhaps his esteem for Mr. Percy's understanding fell some degrees – he considered him as an eccentric person, acting from unaccountable motives.[112]

The morality attributed to eccentricity here is deliberately ambivalent. Percy falls in Oldborough's estimation precisely because he is honest; were he corrupt he could advance himself to a prominent position in public life, which, as far as Oldborough can see, is the only kind of position a gentleman might reasonably aspire to. As a man of the world, Oldborough cannot comprehend Percy's lack of ambition or his love of domesticity; he therefore categorizes him as 'an eccentric person, acting from unaccountable motives'. But, of course, ultimately, the joke is on Oldborough, and the passage satirizes, rather than restates, traditional gender stereotypes.

Lawlessness, Genius and the Cultivation of Eccentricity

Men and women were called 'eccentric' if they were seen to stray beyond their 'proper spheres'. A further, related, use of the astronomical metaphor was derived from the idea that comets were 'lawless' and not governed by any central form of control. In women this was generally associated with irrationality and vulnerability. In men it was more commonly couched in terms of genius.

The novels of, for example, Fanny Burney abound with 'eccentric' female characters whose faculties of reason remain uncultivated, and whose actions are governed solely by passion and imagination.[113] When, in *Evelina* (1778), the eponymous heroine falls in love with Lord Orville, for instance, a concerned Mr Villars tries to make her appreciate her own vulnerability: 'Young, animated, entirely off your guard, and thoughtless of consequences, *imagination* took the reins, and *reason*, slow-paced, though sure-footed, was unequal to a race with so eccentric and flighty a companion.'[114] The 'eccentric' culprit here is imagination personified, who leads Evelina down a flighty and irrational path towards, Villars believes, misguided affections and inevitable misery. Another of Burney's female characters to be disadvantaged by an unrestrained imagination is Mrs Berlinton, in *Camilla* (1796). Brought up by a fanatical maiden aunt who taught her nothing but her faith and prayers, the girl's only enjoyment came from the consumption of 'some common and ill selected novels and romances'. Cut off from society, she believed these books to be accurate representations of the world. 'Nothing steady or rational had been instilled into her mind by others', and so she grew up 'romantic without consciousness, and excentric without intention'.[115]

While later writers tended to equate eccentricity in women primarily with masculinity, and thus with absence of sexual attraction, in Burney's novels, excessive feminine eccentricity is irresistibly attractive to the opposite sex. This is spelled out in *The Wanderer* (1814) by the rakish Ireton in his rejection of the handsome but prudish Juliet. 'Don't take it ill, my love, for you are a devilish fine girl', he concedes. 'But I want something more skittish, more wild, more eccentric ... something that will urge me when I am hippish, without keeping me in order when I am whimsical. Something frisky, flighty, fantastic, – yet panting, blushing, dying with love for me!'[116]

In men, eccentricity as lawlessness was often understood to be symptomatic of genius. 'A strange slightly character!' cries Mr Monckton of Belfield in Burney's *Cecilia* (1782), 'yet of uncommon capacity, and full of genius. Were he less imaginative, wild and eccentric, he has abilities for any station, and might fix and distinguish himself almost where-ever he pleased.'[117] Genius in the Romantic period was conceived of in terms of lawless excess: where ordinary individuals were bound by the rules of artistic or literary composition, for example, geniuses eschewed these rules, creating new ones to replace them.[118] In 1827, Thomas Carlyle restated this Romantic commonplace in astronomical terms: 'Genius has privileges of its own; it selects an orbit for itself; and be this never so eccentric, if it is indeed a celestial orbit, we mere star-gazers must at last compose ourselves; must cease to cavil at it, and begin to observe it, and calculate its laws.'[119]

The 'eccentric genius' was already recognized as a type by the 1780s. For example, in a biographical account of the Quaker Robert Lecky, published in *The Times* in 1787, Lecky's own virtuous character traits are contrasted against those of the 'modern eccentric genius – a heterogeneous character, compounded of absurdity, presumption, and ignorance'.[120] By the turn of the nineteenth century the stock figure of the self-styled eccentric genius had become sufficiently well known to be satirized, as in Edgeworth's *Belinda* (1801):

> Clarence Hervey might have been more than a pleasant young man, if he had not been smitten with the desire of being thought superior in every thing, and of being the most admired person in all companies. He had been early flattered with the idea, that he was a man of genius; and he imagined, that, as such, he was entitled to be imprudent, wild, and eccentric.[121]

The idea that certain individuals, wishing to be taken for men of genius, deliberately affected eccentric manners was explored humorously in Edward Bulwer Lytton's *Pelham; Or, the Adventures of a Gentleman* (1828):

'Look – here', said Glanville, 'are two works, one of poetry – one on the Catholic Question – both dedicated to me. Seymour – my waistcoat. See what it is to furnish a house differently from other people; one becomes a bel esprit, and a Mecaenas, immediately. Believe me, if you are rich enough to afford it, that there is no passport to fame like eccentricity. Seymour – my coat ...'[122]

In advising Pelham how to get on in society, the more experienced Glanville seems to be suggesting that his own public recognition – he has had two books, the subject matter of which is irrelevant for the purposes of the argument, dedicated to him – stems simply from his choice of home furnishings. Eccentricity of a certain kind, he insists, can function as a 'passport to fame'.[123]

The connotations of the term 'eccentric' as applied to fashion and manners were variable. While certain leaders of fashion – such as George 'Beau' Brummel, celebrated for revolutionizing the neck-tie around the turn of the nineteenth century – became staples of eccentric biography, eccentricity of dress was often construed as that realm of originality to be avoided by anyone wishing to appear fashionable. The following quotation, for example, is from George Meredith's *The Ordeal of Richard Feverel* (1859):

That blue buttoned-up frock coat becomes you admirably – and those gloves – and that easy neck-tie. Your style is irreproachable – quite a style of your own! And nothing eccentric. You have the instinct of dress.[124]

Indeed, Pelham, in a series of 'maxims' relating to costume, warns, 'Never in your dress altogether desert that taste which is general. The world considers eccentricity in great things, genius; in small things, folly.' Nevertheless, he takes note of Glanville's advice and, presumably keeping the general taste in mind, adopts what he calls 'the coxcombical and eccentric costume of character'. Within a short while he has the satisfaction of having established 'a certain reputation for eccentricity, coxcombry, and, to my great astonishment, also for talent'.[125] The implication is that it is his apparel, rather than any actual talent which he may or may not possess, that is ultimately responsible for his public estimation.

At some point since the early nineteenth century there has arisen a myth about eccentricity that still has currency today: that to be eccentric one has always had first to be male, to be upper class and to be English. The myth is not true, as this chapter has shown. This is more than just an empirical matter. To assert that the class of eccentric individuals constitutes, and always has constituted, a subset of the male English gentry is implicitly to define eccentricity in terms of belonging. I have argued, conversely, that, eccentricity was defined in terms of not belonging: to be eccentric was precisely to challenge hegemonic classificatory discourses such as those of gender or social class. The social status of individuals of course affected how they were expected to behave. Yet, contrary to what is commonly assumed, in the nineteenth century eccentricity was attributed to individuals from across the social spectrum.[126] The central characters of the case studies that follow – William Martin, Thomas Hawkins and Charles Waterton – came from very different class backgrounds, yet each challenged conventional class strictures in some way.

The historical reasons for the perceived connection between eccentricity and Englishness have yet fully to be explained. Miranda Gill has shown that a discourse of eccentricity permeated many areas of French culture from at least the 1830s, when the word *excentricité* was popularized in the French language.[127] This discourse was closely linked to an 'Anglomania', based on nostalgia for eighteenth-century English aristocratic society, which swept the nation in the wake of the French Revolution.[128] Tellingly, the earliest figurative use of 'excentricité' – by Germaine de Staël in her *Considérations sur la révolution française* of 1817 – pertains directly to the issue of English national character: 'Il n'y a pas de nation où l'on trouve autant d'exemples qu'en Angleterre de ce qu'on apelle l'excentricité, c'est-à-dire une manière d'être tout à fait originale et qui ne compte pour rien l'opinion d'autrui.'[129] Gill suggests that French envy of the greater political liberty and freedom of the press enjoyed by the English may have been a significant factor in the persistence of the stereotype.[130] But it appears that the historical connection between Englishness and eccentricity may previously have been overstated. A sophisticated discourse of eccentricity existed outside England; in addition, I have shown that the English were happy to include foreigners alongside English subjects in eccentric biography. In the case studies, we shall see that when the multifarious people of Britain attended 'eccentric' natural philosophical lectures, bought 'eccentric' pamphlets, read 'eccentric' books, viewed 'eccentric' collections, and visited 'eccentric' gentlemen in their homes, often what they were doing was more subtle and complex than simply admiring their shared national identity.

Over the course of the nineteenth century there appears to have been a shift in moral attitudes towards certain kinds of eccentricity. In volumes of eccentric biography dating from the early part of the century, all manner of licentious, violent, subversive and illegal activities were eagerly narrated: prostitution, cross-dressing, robbery, impersonation. As the century drew on, however, eccentricity tended to be more heavily censored. This shift can be understood as one strand of a broader movement in biographical writing, away from revealing, warts-and-all depictions of biographical subjects towards more sanitized, idealized representations.[131] That the Victorian period can be characterized by a sweeping trend towards conformity, evangelicalism and prudishness is a commonplace, and it comes as little surprise that certain kinds of 'eccentric' activity ceased to be considered fair game for writers of supposedly 'improving' literature. It is sometimes said that the Victorians were 'less tolerant' of eccentricity than their predecessors. A more accurate way of stating the situation might be to say that the boundaries discriminating the 'normal' from the eccentric had been redrawn. As a discourse, eccentricity appears if anything to have become yet more entrenched as the century progressed: eccentric biography flourished; eccentric characters multiplied in novels – the characters of Charles Dickens are probably the best known. The difference was that what passed for 'merely eccentric' – as opposed, say, to mad, illegal or unspeakable – had changed.

It would be misleading to interpret eccentricity solely as a hegemonic discourse functioning to maintain a dominant vision of how society should be ordered. First, the moral stance taken towards many kinds of 'eccentric' activity was ambivalent. 'Eccentric', masculine women, for example, were criticized for their departure from womanly norms of behaviour – they were portrayed as sexually unattractive, unsuited for domestic life, and in some cases improper; however on occasion they were also admired and even celebrated for their courage, fortitude and sharpness of intellect. Within this context, eccentricity possessed subversive capabilities, for while in the courts, for example, fortune vs. poverty, guilt vs. acquittal, even life vs. death could be decided according to an eccentric/not eccentric dichotomy, in everyday life, the divide between permissible and prohibited behaviours was more negotiable. Charlotte Brontë's Shirley, for example, freely describes herself as a manly woman: 'They gave me a man's name; I hold a man's position: it is enough to inspire me with a touch of manhood; and when I see such people as that stately Anglo-Belgian – that Gérard Moore before me, gravely talking to me of business, really I feel quite gentleman-like.' While the people around her are sometimes dismayed by her manner of talking about herself, Shirley is confident and empowered in her masculine roles. She thus potentially serves as an alternative role model for women readers.[132]

Second, to construe eccentricity solely as a label, as something attributed from outside rather than cultivated from within, would be to misrepresent how eccentricity was understood in nineteenth-century British society. As demonstrated by Bulwer Lytton's satirical account of Pelham's manner of dressing, eccentricity could be construed as self-conscious affectation – as a cloak of appearances which could be put on or set aside at will, and which could, if carefully considered, be well received. While instances of self-definition as an 'eccentric character' were rare in the nineteenth century,[133] many 'eccentric' individuals were aware that their activities were unconventional. Like everybody else, they actively sought to manipulate the impressions of those around them through their conduct, and to ignore their agency would be to give, at best, half the picture. The first of the three case studies that follow directly tackles this issue. Focusing closely on the activities of the natural philosopher William Martin and his followers, it argues that 'eccentric' public identities were actively constructed through the dynamic interaction of performers and audiences.[134]

2 PERFORMERS, AUDIENCES AND ECCENTRIC IDENTITIES: WILLIAM MARTIN AND THE WORLD TURNED UPSIDE DOWN

On 5 October 1828 an extraordinary spectacle took place on the North Turnpike Road near Newcastle upon Tyne. A local philosopher took it upon himself to demonstrate to the public a new type of vehicle of his own devising, 'the only Travelling Machine that ever was invented in the World to work from the Centre of Gravity, without the Aid of Horse or Steam'. The philosopher intended the Northumberland Eagle Mail to carry him to the farthest reaches of the kingdom – from England to Scotland, from Scotland to Ireland – to deliver his Lecture on Natural Philosophy wherever a suitable paying audience could be assembled.[1] According to reports in the newspapers, however, he barely survived the short ride to Newcastle from his home town Wallsend.[2]

The Northumberland Eagle Mail was constructed according to the principles of *A New System of Natural Philosophy, on the Principle of Perpetual Motion* (1821). Its designer, also the author of the New System, hoped that its demonstration would secure his position as Northumberland's leading natural philosopher. He also hoped that the people of Northumberland would reward him financially for his ingenuity. Eager to amass a large audience for his performance, he advertised the event in the newspapers, and produced handbills indicating the significance his achievement:

> And as a machanic I have gon as far as a mere mortal can go,
> I can work a carrage from the senter of Gravety, and that does away with the difficulties of hills I now[3]

The promotional campaign was a success. A great crowd assembled on the North Road, mostly on the outskirts of Newcastle, to see for themselves this amazing machine of 'noble and majestic Appearance', which, it was claimed, was a 'superior Mode of Travelling to either Riding or Walking'. Yet setting out from his home in Wallsend, the philosopher had barely got underway when he began to sense 'that his reception was not very flattering'.

The Northumberland Eagle Mail was an extraordinary sight. 'To be brief', commenced the *Durham County Advertiser*, reporting the incident the following week, 'between two immense wheels, on a saddle, which has a prow somewhat resembling an eagle's head, but shaped and projecting confoundedly odd, sat the philosopher. His costume was that of an equestrian'. The vehicle appeared to be an adaptation of the common velocipede, though not obviously an advantageous one:

> A leaning forward and a stroke with both feet together on the ground, (similar to the stroke of a frog in the act of swimming) propelled the machine. Up hill, indeed he progressed backwards like a crab. The labour of the feet by this method, it may be imagined, cannot be called an improvement in walking, as Mr Martin's laborious breathing and flushed countenance fully demonstrated.[4]

The journey to Newcastle took longer than anticipated. The philosopher considered turning back, but he had placed seven gratuity collectors amongst the crowds waiting at the outskirts of the town and he could not bear the thought of their collection boxes remaining empty. Heartened by the encouragement of his friends – whom the *Durham County Advertiser* referred to as 'wags of the press' – he decided to persevere. The decision may have been unwise. By now the 'rabble' had grown 'numerous and impatient'. When the Northumberland Eagle Mail eventually lurched into view, the onlookers vented their frustration by harassing the philosopher, frequently stopping the vehicle in its tracks and once or twice threatening to upset mail, philosopher and all into the gutter. Instead of acclamations of applause and showers of gold, the philosopher was greeted with abuse, ridicule and violence. 'Remonstrance was useless', commented the *Advertiser*, 'and the founder of the new system was forced to dismount and retreat unhonoured and unrewarded – "astonished at the madness of mankind"'.[5]

The demonstration of the Northumberland Eagle Mail was just one of hundreds of performances devised and enacted by the Northumbrian, anti-Newtonian natural philosopher, prophet and poet, William Martin (1773–1851). From the 1820s until his move to London in 1849, Martin was well known in and around Newcastle. '[A] stalwart, good-looking but garrulous eccentric', recalled an elderly writer for the *Northumbrian* magazine in 1883, 'who in 1832 might have been frequently met with in the streets of Newcastle, stick in hand, and carrying sheets of manuscript and printed matter, and maybe a drawing of some new invention, all or any of which he was ready to expound or discuss with the ardour of an enthusiast'.[6] '[A] stout, portly man, perfectly cracked, but harmless', confirmed John Bailey Langhorne, a correspondent of *Notes and Queries*, in 1873. 'He used to strut about the streets very pompously, wearing [a silver medal] round his neck; and was always ready to explain his "philosophy", or his last new invention, and very ingenious he was to any one.'[7] These, like most preserved first-hand accounts of encounters with Martin, were written decades after his death by a nostalgic, ageing generation eager for its memories to be inscribed into local lore: 'Elderly men who have passed their lives on Tyneside must all be more or less familiar with that most singular of beings, William Martin', observed an anonymous contributor to *The Monthly Chronicle of North-country Lore and Legend*, in 1887.[8]

Martin has been handed down by historians of Northumberland as a relatively unproblematic local crackpot. A notable exception is the work of James Gregory, who has considered Martin more analytically in his recent studies of 'local characters'.[9] Martin has not previously attracted the attention of historians of science. Yet his philosophical performances and publications served as the focus for an extraordinary outpouring of creative expression by the people of Newcastle, many of whom were participants and even leaders in the town's scientific culture.

A few days after the demonstration of the Northumberland Eagle Mail, for example, a 'Disciple' of Martin produced a broadside to commemorate the unmitigated success of the event. In sharp contrast to the ridiculous and rather vulnerable figure portrayed by the *Durham County Advertiser*, the Disciple portrayed Martin as a local hero. 'The "Northumbrian Eagle Mail"/ Leaves all coaches at a distance!/ *Time* she keepeth without fail –/ Needs no horses nor assistants'.[10] Over several decades, but especially in the 1820s, hundreds of self-styled disciples and opponents wrangled over the finer points of the Martinian philosophy in person, at tumultuous public lectures and demonstrations, and in print – in broadsides, ballads and the newspapers. This was how he became established, and eventually went down in local history, as a notable 'eccentric' character.

By way of introduction, this chapter begins by outlining some of the key events and relationships in Martin's life. His famous brothers, the historical painter John and Jonathan, incendiary of York Minster, figure prominently. These biographical highlights are followed by a discussion of the philosophical lectures which Martin delivered regularly in inns and other public venues in the Newcastle area. The content of the lectures and the responses of those who attended them are recovered from advertisements, broadsides and other ephemeral print material of the period. Martin's audiences contributed significantly to the performances and this dynamic performer–audience relationship was, I suggest, key to Martin's success.

Martin was violently opposed to the mechanistic, experimental philosophy of Isaac Newton. He also held unorthodox religious beliefs, which were closely tied to his philosophical views. Martin was a marginal figure with respect to both orthodox natural philosophy and the established Church, but his ideas were not entirely idiosyncratic; rather they intersected with well established anti-Newtonian and millenarian, prophetic traditions. The third section explores the notion that some of Martin's contemporaries could have taken his ideas seriously.

But it would not do to take Martin *too* seriously. After all, as the anecdotes cited above begin to show, ambivalence and humour were central to how Martin's 'eccentric' public image was constructed. The final section reads the activities of Martin and his respondents in the spirit of carnival. It suggests that while some contemporaries may have viewed the Martinian philosophy as a genuine threat to the Newtonian world-view, for many more it served as a focus for humorous forms of carnivalesque expression. Thus I argue that, as a local 'eccentric' figure, Martin was symbolically important within his community. Above and beyond providing entertainment, Martinian performance functioned as a medium for communication and polemical exchange on serious matters that were of central importance to Newcastle's political, cultural and scientific life.

William Martin, Natural Philosopher, Prophet and Poet

Martin was born in 1772 in Haltwhistle in Northumberland, the eldest of four brothers and a sister.[11] His family, as he frequently emphasized, was of relatively humble social standing. His father, Fenwick Martin, was successively a tanner, a publican and a coachbuilder. His mother, Isabella, was the daughter of a petty landowner. She was also a religious enthusiast who believed herself to possess visionary powers and, according to Martin's brother Jonathan, she taught her children that 'there was a God to serve and a hell to shun and that all liars and swearers were burnt in Hell with the devil and his angels'.[12] This upbringing appears to have had an important influence on how, as an adult, Martin combined his religious and philosophical views and on how he communicated them to his audiences.

Martin spent much of his childhood living with his maternal grandparents on a small farm in Cantyre, in Argyllshire, Scotland. In the early years of his adult life he worked at a ropery and as a soldier with the Northumberland Militia. In his thirties, however, he turned to inventing, hoping to make his fortune. In 1805, he began to work on the problem of perpetual motion, a pivotal moment in his career. Perpetual motion was no longer a plausible subject for research within elite science. In the eighteenth century, the Royal Society had marginalized the study of perpetual motion, despite, we now know with hindsight, lacking any theoretical basis for believing perpetual motion to be impossible – the science of thermodynamics would not be developed until the 1840s. Nevertheless, the pursuit of perpetual motion was still popular with working-class mechanics and would be for decades to come: as late as 1861, Henry Dircks would complain, in *Perpetuum Mobile; Or, Search for Self-motive Power, during the 17th, 18th, and 19th Centuries*, about the difficulties of arresting the operations of those 'eccentric individuals' who continued to hack away at the 'dry wells and veinless mines of Perpetual Motion'.[13]

Between 1805 and 1806, Martin made thirty-six unsuccessful attempts at constructing a perpetual motion machine, eventually concluding that the task he had set himself was impossible. Then, one night in December 1806, he had a dream which convinced him that he had been chosen by God to solve the puzzle.[14] He made one more attempt and it was a success. The invention consisted of a simple pendulum suspended over an updraft of air. This updraft was created by a pipe concealed under the floorboards, which passively conveyed fresh air from outdoors to a receiver, indoors, that had a small tube poking out of the top. The pendulum was suspended just above the tube.[15] Martin set the pendulum in motion on 4 January 1807. He exhibited it first in Newcastle and then, from January 1808, at 28 The Haymarket, London. Many people paid to come and see it. The most significant visitor, for Martin, was the famous Newcastle-born mathematician Charles Hutton. Hutton had made a bet with his cronies from the Royal Observatory in London that he would be able to work out the secret of the device. He carefully examined it, and on learning that part of it was made of iron, he conjectured that it was done with magnets. Delighted by the error, Martin solemnly told the company that that was not how it was done at all, and Hutton lost his money.[16]

Over the years, Martin made dozens of other inventions, including an inflatable life-preserving jacket, shoes on a new principle and a new version of the miner's safety lamp which he claimed was superior to Humphry Davy's. He was repeatedly frustrated by his failure to gain public recognition for his endeavours; he also complained that many of his best inventions were stolen from him. In 1814, however, he was acknowledged by the Society for the Encouragement of Arts, Manufactures and Commerce, who awarded him its silver medal and ten guineas for his invention of the spring weighing machine.

1814 was the year in which Martin married. He described his wife as an 'inoffensive woman', the same age as himself, forty-two, and a successful dressmaker who had upwards of sixty apprentices during her working life.[17] It was also the year in which he gave up regular employment in order to concentrate on inventing and natural philosophy. Though he had received little formal education, based on his earlier mechanical investigations he began to develop an original system of natural philosophy to rival Sir Isaac Newton's. He published it in 1821 as a small book entitled *A New System of Natural Philosophy, on the Principle of Perpetual Motion; with a Variety of Other Useful Discoveries*. Martin's wife continued to work after the marriage. Dress-making was a lucrative and respectable trade and, on the basis of his wife's income, Martin would have enjoyed a comfortable home and a place in the middling ranks of Newcastle society. Indeed, his wife was probably the main breadwinner of the family. An unreferenced quotation in Richard Welford's *Men of Mark 'twixt Tyne and Tweed* (1895) supports this hypothesis,[18] as does the following extract from a poem written in 1828 by G. W. Couper, a teacher of navigation from South Shields:

> Friend M—n, pray cease to Lecture *for pence*,
> Thy Wife will keep thee as before:
> Thy *Poems* are not *Verse*, and they *Prose* is not *Sense*!
> O M—n, pray Lecture no more![19]

Regardless of the pleas of Couper and other opponents of the New System, Martin devoted the remainder of his life to spreading the word of his philosophy through pamphlets, broadsides and public lectures. Public lectures were just one aspect of the rich intellectual and cultural life of Newcastle in this period.[20] The 1820s saw a concerted effort to establish Newcastle's position as a leading centre for the arts and sciences, with the foundation of the Society of Antiquaries of Newcastle upon Tyne (1813), the Northumberland Institution for the Promotion of Fine Arts (1822), the Northern Academy of Arts (1828) and the Natural History Society of Northumberland, Durham and Newcastle upon Tyne (1829). The Newcastle Literary and Philosophical Society, founded in 1793, was relocated to impressive new premises in 1825; courses on science and other subjects were offered regularly, by itinerant lecturers and by the founder Secretary of the Society, the dissenting minister William Turner. An alternative, more affordable venue for lectures was the Newcastle Literary, Scientific and Mechanics' Institution, founded by the radical printer Eneas MacKenzie, in 1824. Around this time, Mechanics' Institutes were being set up across the country by Whig and radical reformers, with the aim of educating members of the lowest social classes in the hope that this would ultimately lead to social reform.[21] Martin did not, in fact, lecture at the Literary and Philosophical Society or the Mechanics' Institution – indeed he was vehemently opposed by key senior members of both, something which will be explored in detail later on. But he did succeed in drawing large, receptive audiences to performances at the Newcastle Theatre Royal and at inns which, in addition to functioning as meeting places for councils, societies and clubs in this period, commonly provided entertainments such as lectures, musical performances and exhibitions of paintings, panoramas, automata and other curiosities.[22]

After his wife's death in 1832, Martin's fortunes declined, but he retained his presence as a prominent local character. In 1845 the painter and poet William Bell Scott was introduced to him in the street by his friend Captain Weatherley. He reports that Martin, 'with exaggerated politeness, drew his feet together, bent forward, lifted his tortoiseshell hat high in the air, and answered "Gratified to meet you sir! I am the philosophical conqueror of all nations, that is what I am! and this is my badge."' When he unbuttoned his jacket to reveal a medal the size of a saucer, Scott was unsure how to react. His conclusion was that Martin was 'manifestly crazed', yet he had an air about him that commanded respect. 'A noble presence even was his', he wrote, 'although he was poor enough to sell his pamphlets thus on the street.'[23] By now, Martin was in his seventies and his livelihood was waning. He died six years later, in 1851, having spent the last months of his life living in London at the home of his brother John.

Martin's brothers exerted considerable influence on him throughout his life. Two of them, John and Jonathan, were well-known public figures in their own right. As such, they informed how Martin was viewed by his contemporaries. They also figure prominently in most posthumous accounts of Martin's life.

John, the youngest brother, was a highly-successful painter.[24] Having been first apprenticed to a coach-painter in Newcastle, he moved to London in 1805 and from 1811 exhibited paintings at the Royal Academy and the British Institution. By the 1820s, he had become extremely popular with the exhibition-going public for his dramatic paintings on classical and biblical themes: *The Fall of Babylon* (1819), *The Destruction of Herculaneum and Pompeii* (1822) and *The Seventh Plague of Egypt* (1823), for example. His distinctive style, which became known as the apocalyptic sublime, was characterized by architectural landscapes on a vast scale, dramatic contrasts between darkness and light, and fiery imagery inspired by secular and sacred history. In *Belshazzar's Feast* (1820), for example, the writing on the wall blazes down ominously upon the horrified company as the prophet Daniel, standing in the centre of the composition, interprets the cryptic signs as foretelling the end of Belshazzar's kingdom (see Figure 2.1). John

Figure 2.1. *Belshazzar's Feast* (1820) by John Martin. As Belshazzar, King of Babylon, and his guests drank wine from the vessels which his father Nebuchadnezzar had taken from the temple at Jerusalem, they praised the gods of gold, silver, bronze, iron, wood and stone. Because of this profanity, a hand appeared and began writing on the wall. The Laing Art Gallery, Tyne & Wear Museums.

Martin made a sizeable fortune exhibiting his paintings to paying audiences and selling engravings of them. He also illustrated books, ranging from a new edition of *Paradise Lost* (1827) to the palaeontologist Gideon Mantell's *Wonders of Geology* (1838), for which he drew the frontispiece.[25]

John Martin moved in elevated social circles and associated with some of the great artistic, literary and philosophical talents of his day. Samuel Carter Hall recalls that he gave weekly parties at his fashionable London home, 'at which many men of celebrity, and others who were commencing lives that afterwards became celebrated, were guests'.[26] Many of his close associates were political reformers, and it has been suggested that paintings such as *Belshazzar's Feast* were meant to convey to discerning viewers 'the folly of grandeur, the inevitability of punishment following hubris'.[27] John held rationalist and deist views: he once challenged his friend Ralph Thomas: 'Why should a man use one kind of logic for religion and a different kind for general affairs?' Thomas was even more shocked when John admitted to believing, based on geological evidence, that the Earth was many millions of years old and was peopled with inferior animals before the time of man.[28] On this particular point, his brother William disagreed with him, preferring to interpret the scriptural account of Creation literally. Nevertheless, John would have been an important source of new ideas in politics, religion and science. Moreover, his visionary paintings provided inspiration for the Northumbrian philosopher as he constructed his own prophetic persona.

Jonathan Martin's life story is more tragic.[29] As a young man, he was press-ganged into the Royal Navy. In 1810, having been paid off from the Navy, he married and soon after, despite his wife's objections, began to frequent Methodist meetings. Following a sudden conversion at a prayer meeting, he became very hostile to the Church of England and began to disrupt church services. In 1817, he threatened to shoot the Bishop of Oxford and was committed to a lunatic asylum. In 1821, a few months after the death of his first wife, Jonathan escaped. He included a dramatic account of the adventure in *The Life of Jonathan Martin, of Darlington, Tanner: Written by Himself*, etc. (1825), which included details of fantastical, portentous dreams. For several years, he preached and made his living working as a tanner and hawking his *Life* – he claimed to have sold around 14,000 copies by 1828.[30] That same year he moved to York with his second wife and on 1 February 1829, at God's command, he set fire to York Minster causing £100,000 worth of damage.

In the weeks leading up to his trial, Jonathan became a celebrity. He had his portrait painted,[31] and hundreds of visitors were admitted to look at him through the window of his cell in York prison. The most prestigious were allowed to meet him in person. On 28 March, the *Yorkshire Gazette* reported:

Noblemen and titled ladies, a crowd of persons of rank and distinction, throng to Martin's levees; they are all very graciously received, have the honour to shake the incendiary's hand, and depart highly gratified. [Jonathan] Martin, on his part, is no less flattered, and declares that he never in his life shook hands with so many people of quality.[32]

At his trial, when asked whether he was guilty or not guilty of setting fire to the Cathedral Church of Saint Peter of York, Jonathan answered, 'It was not me, my Lord, but my God did it.' He was declared not guilty on the grounds of insanity and was taken to Bedlam, the Criminal Lunatic Asylum in Lambeth, London, where he remained until his death nine years later. In 1833 an anonymous writer for Tilt's *Monthly Magazine*, reporting on an excursion to Bedlam, observed that 'the York Minster incendiary ... had monopolized the visiting interest'.[33]

William respected Jonathan's zeal. He shared his brother's intense religious feeling and, like Jonathan, believed himself to be in direct communion with God, experiencing prophetic dreams and interpreting passages of Scripture as references to his own life. He included a celebratory poem 'On the Burning of York Minster' in his own autobiography, *A Short Outline of the Philosopher's Life, From Being a Child in Frocks to This Present Day, After the Defeat of All Impostors, False Philosophers, since the Creation; By the Will of the Mighty God of the Universe, He Has Laid the Grand Foundation for Church Reform by True Philosophy &c.* (1833). The following year, in *The Christian Philosopher's Explanation of the General Deluge, and the Proper Cause of All the Different Strata: Wherein It Is Clearly Demonstrated, That One Deluge Was the Cause of the Whole, Which Divinely Proves That God Is Not a Liar, but That the Bible Is Strictly True* (1834), he boasted that Jonathan could 'confound either bishop or any priest with the holy scriptures; he knows, as well as myself, that they are all corrupted'.[34] Martin did not see eye to eye with his brothers on every aspect of natural philosophy or religion. But family was centrally important to him, both in terms of the development of his own ideas and in terms of how he represented himself to the public.

Despite their humble origins, he argued, the brothers had all shown themselves to be men of great talent. John was 'the greatest historical painter, such as the world never knew'; Jonathan was 'well known throughout the kingdom'; the fourth brother, Richard, a soldier, was 'a man of great talents, although not a public character'; and Martin himself, thanks to continual divine assistance, was 'the greatest philosopher'.[35] In line with his mother's prophesy that one day her sons' names would sound from pole to pole,[36] Martin believed that all four brothers had been sent by God as a lesson to 'Infidels and Impostors', and he did his best to make this known as widely as possible.

The Lecturer and his Audiences

Martin first felt compelled to lecture on natural philosophy in the spring of 1827. 'When the spirit first pressed on my mind to go forth and lecture against their false systems', he later recalled, 'I thought that, not being a good speaker, I should not be able to do it; but then it struck me that Moses was the same, and yet God raised him up to do wonders by his power.' Finding strength in the tenth verse of the forty-first book of Isaiah, 'Fear thou not, for I am with thee; yea, I will help thee; yea, I will uphold thee with the right hand of my righteousness', Martin 'became bold, and not afraid of all the world'.[37] He announced his intentions in broadsides and newspaper advertisements (see foot of Figure 2.2):

> WILLIAM MARTIN, the NATURAL PHILOSOPHER, is now going to visit all the principal Towns in the United Kingdom, and give one Lecture to sow the Seed of Truth, so that it may grow to the End of the World, for the Good of all Mankind, to vanquish Infidels and Impostors; and those that miss this important Visit, will never be indulged with the Philosopher, but his Disciples, – for he must do the Mission that the God of Heaven has appointed him to do.[38]

The first lecture was arranged for Wednesday 4 April 1827. According to a letter to the editor of the *Tyne Mercury* the following week, at a quarter past seven, Martin took the chair before a select but respectable audience at the Dun Cow, Quayside, Newcastle.[39] The letter to the *Tyne Mercury* gives some details of the contents of Martin's first lecture; others can be deduced from his *New System of Philosophy*, for his advertisements show that he always gave the same lecture, and that this followed the outline of his book (see Figure 2.2).

The Dun Cow lecture, like all subsequent ones, began with Genesis 2:7, 'And the Lord God formed man of the dust of the ground, and breathed into his nostrils the breath of life, and man became a living soul.' This verse, Martin explains in the *New System of Philosophy*, 'completely proves, that AIR is the real Cause of the Perpetual Motion; without which neither man nor beast can live, vegetation nor the fishes of the sea exist'.[40] The word of God, and Martin's invention of 1807, together demonstrated the fundamental principle of the Martinian system of philosophy that, after God, air was the 'the cause of *all things*'.[41] Martin then demonstrated how this principle could be applied to different philosophical problems. The tides, for example, Sir Isaac Newton had erroneously attributed to the gravitational pull of the Moon. In fact, tides were caused by air. Imagine a pea rested on the end of the shaft of a pipe, Martin instructed his listeners; blow gently on it and you will see that it rotates and oscillates back and forth. The planets, he argued, are suspended in the same way in air meaning, significantly, that each planet has not one motion, but two: a rotational one and an oscillating one. According to the Martinian system, the daily tides are caused not by the moon, but by the oscillating motion of the Earth: 1,348 oscillations in one

MARTIN,

Natural Philosopher!

Patronised by his Grace the Duke of Northumberland.

Ladies and Gentlemen, I beg Leave to inform you, that the PHILOSOPHER MARTIN delivered his first Lecture on the MARTINIAN PHILOSOPHY founded on the Principle of Perpetual Motion, in a spacious Room, at the Dun Cow, on the Quayside, Newcastle, on the Evening of Wednesday, the 4th of April, 1827, to a select but respectable and highly gratified Audience, amidst the most lively Acclamations of Applause.

POETRY,

On the Martinian System of Philosophy, by a Friend and great Admirer of the New System.

O Martin! O Martin! how vague and uncertain
Have Systems been until now!
But thy Master-mind, as the learned will find,
Set Philosophy right at the Cow!
Sir Isaac himself thou hast laid on the Shelf—
His Works rendered merely waste Paper;
For thy golden Rule has proved him a Fool,
And extinguish'd him just like a Taper!
He nothing could tell of Heaven or Hell,
Of Celestial Things he'd no Notion;
Whilst Martin has trac'd where each of them's plac'd,
By the Aid of Perpetual Motion!
He has them both seen—in a Vision I mean—
But what could we wish for more certain,
Than to have that reveal'd, which before was conceal'd,
By the Vision of Professor Martin?
The Size of the Sun—What in it is done—
For there is the Heaven of Heavens!
He frees from all Doubt, with more Things, about
Which we've long been at Sixes and Sevens.

The Tides' ebb and flow—the World's overthrow—
The beautiful Planets' Vibration—
The Ice at the Poles—the Nature of Souls—
And the Vision's Interpretation—
The Nature of Comets—the wand'ring Planets—
The Size of the Universe taken
To the Breadth of a Hair, with Remarks upon Air
From the System, if properly shaken.
But the Moon, it is clear, is his favourite Sphere,
And lunar Affairs his Vocation;
For in Prose or in Poem he's here more at Home,
Than any where else in Creation.
Then proceed mighty Man in maturing thy Plan,
The Children of Men to enlighten;
Till all the learn'd Bats to their Holes, like the Rats,
With the Blaze of thy Genius thou frighten.
Publish all thou hast penn'd, from Beginning to End,
And give all the Gonnerels a Smarting,
With thy Cat-o-nine-Tails, which we know seldom fails,
To cut up the Snarlers at Martin.

THE CONTENTS OF THE LECTURE.

1. A new System of Philosophy, upon the Principle of Perpetual Motion.
2. The Magnitude of the Sun ascertained.
3. Observations on the Moon and the Earth.
4. The Sun's Vibration on the Earth.
5. Observations on the Ice breaking away from the North Pole, and on the fixed Stars.
6. The Vision in 1806, and the Interpretation.
7. Concerning the Destruction of the World.
8. The Explanation of Perpetual Motion.
9. Observations on the Irregular Velocity of the Tides on all Coasts where the Tides ebb and flow.
10. General Observations on Air.
11. Explaining the Sun to be the Heaven of Heavens.
12. Explaining the Nature of Starry Heavens.
13. Explaining the Nature of Comets.
14. Explaining the Nature of the Wandering Planets.
15. Pointing out the Size of the Universe.
16. Concluding with the Philosopher's Ideas of the beautiful Planet the Sun,—the Heaven of Heavens.

WILLIAM MARTIN, the NATURAL PHILOSOPHER, is now going to visit all the principal Towns in the United Kingdom, and give one Lecture, to sow the Seed of Truth, so that it may grow to the End of the World, for the Good of all Mankind, to vanquish Infidels and Imposters; and those that miss this important Visit, will never be indulged with the Philosopher, but his Disciples,—for he must do the Mission that the God of Heaven has appointed him to do.

Tickets of Admission 2s. each.　　　　　WILLIAM MARTIN, *Natural Philosopher.*

NEWCASTLE: PRINTED BY EDWARD WALKER.

Figure 2.2. Martin was an acute self-publicist. He once insisted that before lecturing at the Newcastle Theatre Royal, 'It must be in all the 3 papers and a thousand handbills over the town/ And that must continue a fortnight before I am calld Northds honour to crown.' Reproduced by kind permission of the Syndics of Cambridge University Library.

year resulting in 1,348 tides.[42] In a similar way, Martin continued, the oscillations of the Sun are responsible for the alternation between summer and winter on Earth. From these two facts, the relative sizes of the Earth and Sun can be calculated. If the Earth oscillates 1,348 times in a year and the Sun only once, this must mean that the Sun is 1,348 times the size of the Earth. Sir Isaac Newton, being unacquainted with the powerful influence of perpetual motion on the planetary system, computed the magnitude of the Sun to exceed that of the Earth by 900,000 times. Newton, Martin scoffed, was a fool, for on this matter he was 898,651 times wide of the truth![43]

Martin continued the astronomical theme in the following sections of the lecture, with observations on the Moon, Earth and fixed stars. In the sixth section, he related the portentous dream of 1806 which had convinced him that he was chosen by God to discover the perpetual motion. The dream began with an eclipse of the Sun, representing, Martin explained, the intellectual darkness – the dominance of the Newtonian philosophy – which loomed over Britain prior to the discovery of perpetual motion. Martin was standing naked in a vast forest – his nakedness symbolizing his poverty, and the forest symbolizing the Royal Navy, to whom he had sent details of many useful inventions – when suddenly he beheld a great lion, as high as a horse, with limbs as strong as those of an elephant, a large shaggy mane and a fierce countenance. He turned to run but the lion was soon within a few yards of him. Just as he had given up all hope of escaping, it ran straight past him at an incredible pace. After it had run a further fifty yards or so, it suddenly stopped and turned to face him; its ferocious countenance softened and it now walked by Martin's side, licking his hand and looking him pleasantly in the face as a faithful dog looks up to his master. The lion, Martin explained, represented the British Government, which passed him unheard for many years but which would, at some future time, take notice of him and his inventions.

In the remaining sections, Martin delivered observations on the Sun, the starry heavens, comets, wandering planets and the size of the universe. The lecture concluded with some ideas about the Sun being the Heaven of Heavens. At precisely fifteen minutes before nine he stopped talking and the patrons of the Dun Cow rewarded him with lively applause.

Encouraged by the success of the Dun Cow performance, Martin resolved to repeat his lecture at as many venues as possible. Throughout 1827 and 1828, he placed regular advertisements for his lectures in the Tory *Newcastle Courant*. He offered to lecture to rich and poor alike. For ladies and gentlemen, he offered to travel to any distant part of the county provided that forty or more subscribers could be recruited to pay five shillings each. Tradesmen in the neighbouring counties could hear the lecture for two shillings provided a company of no less than a hundred could be assembled. For the benefit of working men, he offered to lecture at public institutions, where a collection would be made, half to go to the lecturer and half to be kept by the institution. He even offered to travel to London to deliver his lecture in person to King George IV should he desire to hear it.[44]

Drawing creatively on advertising techniques developed during the late eighteenth century, Martin used poems and lively narratives in his advertisements in an attempt to attract new audiences and to imprint his name 'William Martin, Natural Philosopher' and the name of his product, the 'New System of Philosophy', onto the memories of the people of Newcastle.[45] Helpfully, many of his advertisements include elaborate accounts of previous lectures and thus give a reasonable idea of who attended and what happened. Audiences comprised members of both sexes from diverse social classes, ranging from gentlefolk to tradespeople: Martin's advertisements mention a vicar, a butcher, a shop-keeper, a doctor, a mathematician and a soldier amongst others. Although in practice many working people may have been excluded by the one shilling admission fee, the lectures functioned as events at which members of different social classes could mingle: in an advertisement dated 22 May 1828, for example, Martin thanks 'the Ladies and Gentlemen of Blanchard, for their numerous Attendance and Politeness; and also the cheerful Lead-miners for their manly and genteel Conduct'.[46]

Martin claimed to have lectured to groups of upwards of 300 people on occasion.[47] Sometimes he was unprepared to cope with the extent of his own popularity. At Ulgham on 5 June 1828, so many people turned up that they could not be fitted into the Sun Inn and the lecture had to be moved to a nearby barn; the following week more than forty people had to be turned away from the George and Dragon at Alnwick.[48] It is impossible to calculate precisely how many people saw Martin lecture in the period 1827–8, especially as Martin may well have 'puffed' his lectures somewhat in his advertisements; however, at the height of his career, Martin was lecturing several times a week and, taking 300 as an upper limit, I would estimate the figure to be of the order of 10,000, possibly including many repeat visits by dedicated 'disciples'.

Martin's advertisements also give an indication of how his lectures were received. Audience behaviour, it seems, was unpredictable. Sometimes the lecture went down very well. In an advertisement of 5 April 1828, for example, he thanked the ladies and gentlemen of Hetton-le-Hole for 'so numerously' attending his lecture at Mr Billsborrow's, and for honouring him with '3 times 3' (three rounds of three cheers); he also thanked the attendees of his Ovingham lecture, some of whom had walked nine miles to attend, but who were most gratified and convinced that the New System was not to be overturned.[49] On 2 May, patrons of the White Swan at Morpeth drank 'Success to the new System' and again rewarded the lecturer with cheers. One 'Gentleman of Morpeth, a Disciple of Mr M.'s' penned a verse in honour of the occasion. Martin inserted it into his next advertisement:

> Hail! mighty Martin! man of master mind!
> Born to instruct and to improve mankind;
> Bound by no ancient, no pedantic rules,
> Has proved Boyle, Bacon, and old Newton, fools.[50]

Martin's reception at the Black Horse at Bishop Aukland on 26 May was even more convivial. He was welcomed by a band who played with cheerful glee, and at the end of the lecture the company honoured him with 3 times 3, named him 'Champion of true Philosophy', drank 'Success to the new System, as being the only one on Earth calculated to enlighten the People who have their proper Senses', and gave him strict orders to repeat his visit as soon as possible.[51]

Martin succeeded in amassing a band of supporters who styled themselves Martinian 'disciples'. These individuals assisted Martin financially: in an advertisement of 7 June 1828, for example, he noted two gentlemen disciples of Monkseaton who 'were very liberal in rewarding the philosopher'.[52] They also expressed their devotion in doggerel verse. During 1827 and 1828, poetical declamations of support for Martin and his philosophy erupted from the presses of Northumberland by the dozen; they were circulated in the newspapers and as single-sheet ballads, sold for a penny or sometimes given gratis at the writer's expense. In May 1828, for example, 'A Disciple' of Morpeth produced a poem in response to Martin's lecture at Henry Esther's:

> GREAT MARTIN! as the Apostles did,
> Derived his lore from Heaven;
> By him Sir Isaac's followers now
> Clean off the field are driven.[53]

And just a week later a certain 'F.C.' composed a poem which he recited to the assembled company after a lecture at Usworth. He afterwards had it printed up as a broadside. The poem began:

> Hail! Martin Hail! bright star of Genius Hail!
> Thy light and easy system must prevail;
> And sage Philosophers we shall be,
> If but obedient to be taught by thee,
>   Great Martin![54]

Everyone was delighted, and when the poem was finished Martin remarked, 'Aye he's one of my disciples, he's one of my disciples, I have hundreds of them up and down the Country.'[55]

Audiences were not always so well behaved however, as Martin readily admitted in his advertisements, and troublemakers soon became a regular feature of the lectures. Some attempted to defeat the philosopher by proposing counterarguments to the New System, while others preferred to disrupt the lectures by heckling and creating distractions. On one occasion in January 1828 at the Commercial Hotel at North Shields, for instance, a minister who could not bear to hear Newton criticized tore up his admission ticket before Martin's eyes.[56] When Martin repeated his lecture there in March, an attempt was made by 'a notoriously *drunken* Cobler' and 'a little *hairy Frenchman*' to disturb the assembly, but they were soon drowned out by loud cheers.[57] At Stamfordham there was a self-opinionated man 'who curled up his Nose like a wild Ass snuffing the Gale' at Martin's exposure of the Newtonian philosophy.[58] At Ponteland there was a butcher who, not liking Martin's explanation of Hell, ran in and out of the room 'like a Hen with Egg'; there was also a Newtonian reverend gentleman who pestered Martin about what instruments he used to prove the size of the Sun but was defeated when Martin asked him what instruments Sir Isaac had used to the same end.[59] At Sunderland, Martin met with yet another irritable individual, this time a young shopkeeper, who was 'much put about on hearing the old System cut up'. He seems to have been bent on there being some third cause, in addition to the two Martin identified (the deity being the first great Cause, and air the second according to the Martinian philosophy). Several 'Friends of the new System' challenged the troublemaker to name the third Cause but, Martin recalls with satisfaction, 'he could make no proper Answer, but threatened to knock the People, who maintained the new System to be right, through the Window'. Thankfully, 'nothing else particular occurred'.[60]

Audience behaviour at Martin's lectures became increasingly outrageous. After a lecture at Causey Park Bridge in May 1828, for example, the following occurred:

> a little Man, whom the Philosopher denominates Tom Thumb, was encouraged by a silly Doctor to make a Speech in Behalf of the old System, which ended with Tommy showing the People how he could stand on the Crown of his Head upon the Table; – another Instance that none but Half-wits ever attempt to overturn the new System.[61]

Another ridiculous situation arose at Whickham. A 'silly' landlord, encouraged by 'a complete Ignoramus of a Doctor' who whispered in one ear, and 'a Halfwit' who was busy on the other, began, 'through his Ignorance, to talk Nonsense'. By strange coincidence a Newfoundland dog happened to come into the room at that moment in search of his master. Being gratified on finding him, Martin recounted a few days later, 'the noble Animal sallied behind the Goneral Trio, as they were making the most Noise, and began to bark' ('goneral' was one of Martin's favourite insults, presumably meaning something like 'scoundrel'). Martin turned to his audience and wittily commented that it was surely time to be done, when the dog was barking at the trio's foolishness. The riposte 'produced much Laughter and good Humour'.[62]

Martin's audiences were extremely active. Audience members cheered and applauded and jostled and threatened and argued and barked and stood on their heads, and every such act contributed to the performance as a whole. Martin delivered essentially the same lecture dozens and dozens of times, yet each lecture was unique, for its meaning depended as much on the contributions of audience members as on those of the lecturer himself.

Sometimes Martin's opponents chose to air their grievances in print. In some cases, full-scale pamphleteering battles ensued, with Martin issuing counterattacks, always at a price, and disciples and adversaries fighting it out in the streets through the medium of print. In June 1827, for example, Martin produced a ballad entitled 'The Downfall of the Newtonian System' (see Figure 2.3). The ballad was an attack on an 'Astronomer' of Newcastle, accusing him of purposely misleading the public by lecturing on the Newtonian philosophy. Although Martin did not name the 'Newcastle Astronomer' directly, the ballad was known to be aimed at the local mathematician and schoolteacher Henry Atkinson.[63] Within a matter of days, a broadside headed 'Ecce Homo' and authored under the pseudonym 'A Newtonian' appeared in response (see Figure 2.4). A defence of 'Mr H. A.', 'Ecce Homo' aimed to undermine Martin's credibility as a philosopher by showing him to be 'ignorant of the most simple rules of orthography'. The writer drew attention to Martin's more amusing gaffes: 'he calls himself a "CUNT–RY JOHNY"', he scoffed, inserting an extraneous hyphen to hammer the joke home. 'Now, by the Soul of Rabelais, this is "for Mirth too much;" but, Reader, spend a Penny in the Purchase of this most *delectable Morceau*, and, if thou art not a Stoic, thou wilt laugh until thy Sides ache, or commiserate his Condition ... As this INCORRIGIBLE IGNORAMUS is, by his Works, proved to be "Mad! madder than the maddest of March hares!"'[64]

Figure 2.3. Martin's broadsides often featured decorative borders, which made them more attractive to potential purchasers. Reproduced by kind permission of the Syndics of Cambridge University Library.

Ecce Homo !

"In whatever Way a Man becomes notorious, he is equally an Object of public Curiosity."

THOSE who have had the Management of the self-dubbed NATURAL PHILOSO-PHER's "Rhyme run mad," just published, have done an essential Service to the Public by shewing him up in his TRUE COLOURS. Hitherto the most *unwarrantable Liberties* have, by CERTAIN TYPOGRAPHERS, been taken with his *brainless Productions, by putting them into Language ;* not so with this last, as it is strictly *verbatim et literatum.* Here is a Fellow attacking a most respectable Teacher (Mr H. A.) and pretending to lecture upon Philosophy, who knows no more about that Science than " MR TEASDALE's DUN COW," or " THE HEAD OF THE NAG," and who is at the same time ignorant of the most simple Rules of Orthography ; as, for Instance, speaking of his own great self Acquirements surpassing the combined Knowledge of all the " Flats" of Oxford and Cambridge, he calls himself a " CUNT-RY JOHNY." Now, by the Soul of Rabelais, this is " for Mirth too much ;" but, Reader, spend a Penny in the Purchase of this most *delectable Morceau,* and, if thou art not a Stoic, thou wilt laugh until thy Sides ache, or commiserate his Condition, and exclaim

" Alas !
" Poor gentleman, I am sorry for ye ;
" And pity much your *upper story."*

As this INCORRIGIBLE IGNORAMUS is, by his Works, proved to be

" Mad ! madder than the maddest of March Hares !''

is it not dangerous for the lives of his Majesty's Subjects to be exposed

" too near a man
" In so Peg Nicholson* a situation" ?

A NEWTONIAN.

Newcastle, June, 1827.

* Margaret Nicholson, a Maniac, who, in 1786, made an attempt on the Life of his late Majesty.

T. Blagburn. Printer. Old Flesh-Market.

Figure 2.4. In the opening quote, the author of this broadside attacking Martin – probably the schoolteacher Henry Atkinson – satirizes the pervasive, and in his opinion pernicious, public preoccupation with fame and notoriety. Reproduced by kind permission of the Syndics of Cambridge University Library.

It was only a matter of days before Martin retaliated. In a pamphlet entitled 'Lines on the Safety Lamp, with a Reply to an Attack Headed "Ecce Homo"', he turned on the perpetrator of the attack, threatening legal action on the grounds of libel. Careful not to commit a libel himself, he cleverly insinuated, without insisting, that Henry Atkinson had produced the broadside himself:

Sume says it's the Newcastle Astronomer that has done this infamus thing
But if it is the case his honour is fast taking great wing.[65]

Finally, Martin's disciples got in on the action too (Figure 2.5). In an anonymous broadside entitled 'True Philosophy!' a friend of the New System praised the clarity, truth and close scriptural accordance of Martin's works, and questioned the sanity of those who failed to see their brilliance:

> And the 'Newtonian' should be purg'd and bled,
> His head clean shav'd – then quietly put to bed.[66]

To Martin and his disciples, if anyone was 'Mad! madder than the maddest of March hares', it was the Newtonians, of course.

Newton was Martin's nemesis. In dozens of the pamphlets and broadsides which he published after the *New System*, as well as in private correspondence, he signed himself 'William Martin, anti-Newtonian'. His chief issue with the Newtonian system was that, for all its clever mathematics, it described only *effects*; never did it show the *cause* of phenomena, which in Martin's eyes was the true goal of natural philosophy. Newton's first error, according to Martin, had been to place too much faith in his own observations and experiments, and not enough in the Word of God: his hubris was the reason for his failure. Martin's challenges to the dominance of the Newtonian, experimental philosophy were amongst the reasons that he was labelled eccentric. And yet, as I show in the next section, his beliefs and methods would not have appeared as inherently nonsensical to contemporaries as they perhaps do to the twenty-first-century observer.

TRUE PHILOSOPHY!

——

ALL you who true Philosophy would learn,
Read Martin's Works, and there you will discern,
" Plain as a pike-staff, or as A, B, C,"
How well with Sacred Writ the Writer doth agree.

As to the author of " Mandate Imperial,"
He shews deficiency of brain material ;
And the " Newtonian" should be purg'd and bled,
His head clean shav'd—then quietly put to bed.
There's none so craz'd as these in all the race of
      Adam—
Not even BILLY SCOTT, or DOODIMDADDIM !

Proceed, great MARTIN, with thy grand design!
Ne'er fear the snarling Curs of Coaly Tyne.

Figure 2.5. The author of this broadside, a disciple of Martin, unfavourably compares supporters of the Newtonian world-view with Billy Scott and Doodimdaddim, both of whom feature in H. P. Parker's *Eccentrics and Well Known Characters of Newcastle* (see Figure 2.8). Reproduced by kind permission of the Syndics of Cambridge University Library.

The Martinian Philosophy in Context: Anti-Newtonian and Millenarian Traditions

Martin was not alone in his opposition to Newton. It is well known, for example, that many of those now associated with the Romantic movement expressed hostility towards the mechanistic view of nature that they saw as being central to Newtonian science. As John Keats famously wrote in *Lamia* in 1819:

> Do not all charms fly
> At the mere touch of cold philosophy?
> There was an awful rainbow once in heaven:
> We know her woof, her texture; she is given
> In the dull catalogue of common things.
> Philosophy will clip an Angel's wings,
> Conquer all mysteries by rule and line
> Empty the haunted air, and gnomed mine –
> Unweave a rainbow.[67]

More appealing to the Romantics were the natural philosophies of the German polymath Johann Wolfgang von Goethe and the German Idealists, such as Friedrich Schelling, whose *Naturphilosophie* rested on an organic, dynamic vision of nature rather than on an atomist, mechanist world-view.[68] The Romantic conception of 'genius' valued personal insight and individual expression over cold rationality. Samuel Taylor Coleridge, for example, wrote to his friend Thomas Poole, 'I believe the souls of 500 Sir Isaac Newtons would go to the making up of a Shakspere or a Milton.'[69] While Martin himself cannot usefully be labelled 'a Romantic', it appears that aspects of his philosophy may have been drawn from Romantic sources, possibly under the influence of his brother John, the painter. Moreover, many of his respondents in 1820s Newcastle would have been familiar with anti-Newtonian sentiments as expressed by some of the influential thinkers of the day.

For example, Goethe's theory of colour is one strand of Romantic, anti-Newtonian philosophy which Martin appears to have drawn upon. Goethe approached the study of colour from the position of wishing to understand how artists could use colour to the best aesthetic effect, and in 1810 he published a treatise *Zur Farbenlehre* in which he flatly rejected Newton's account. He objected to Newton's mathematical approach to the study of colour because it could not take account of the subjective nature of perception. Moreover, he disagreed with Newton's theory that white light was a compound of coloured rays, arguing instead that white light was homogenous and that colours were produced as a result of the interactions between darkness and light.[70] For example, Goethe explained the creation of the colour blue thus: 'If … darkness is seen through a semi-transparent medium, which is itself illuminated by a light striking on it, a blue colour appears: it becomes lighter and paler as the density of the medium is increased.'[71] Goethe's theory of colour was based partly on observation, but it also had a moral justification in that it restored light's unified, harmonious character. One artist who engaged very closely with the theory was J. M. W. Turner, who produced studies with titles such as *Light and Colour (Goethe's Theory) – the Morning after the Deluge – Moses Writing the Book of Genesis*, which was exhibited in 1843.[72] John Martin's paintings, with their brilliantly contrasting areas of light and darkness, may also have been influenced by Goethe's theory (John Martin knew Turner), and it is possible that William Martin may have learned about the theory from his brother. Although Martin did not write at length on the subject of optics, his explanation of the 'beautiful azure colour' of the sky in the *New System* bears a remarkable similarity to Goethe's:

> As the glorious Sun scattereth its rays in all directions, until they are lost in the immensity of distance, and the regions of darkness commence, that pitchy darkness, forming, as it were, a back-ground to the regions of light, and, in a manner, seen through that transparent medium, will naturally communicate to the sky an azure tint of the most transcendent beauty, as we now behold it.[73]

Caution is therefore required when connecting Martin's anti-Newtonianism with his perceived eccentricity. J. M. W. Turner may have been considered eccentric for other reasons – his miserable physical appearance was often commented upon – but the way in which he handled colour, in theory and in practice, was partly why he was celebrated as one of the great artists of his time. Challenging the theories of Newton did not automatically make one an eccentric character in this period.

Other anti-Newtonian traditions may have had an even greater influence on Martin's natural philosophy and on how it was received. The Muggletonians, for example, were the plebeian followers of the seventeenth-century prophets John Reeve and Lodowick Muggleton. Scattered throughout Britain, they never numbered more than a few hundred; for the most part they had little impact on those outside their sect as they did not tend actively to promote their beliefs. In the early nineteenth century, however, in response to the explosion of lecturing on astronomical topics in Mechanics' Institutes, Literary and Philosophical Societies and other public venues, the Muggletonians began to speak out in opposition to orthodox 'Newtonian' natural philosophy.[74] Based on their literalist, common-sense readings of the Old Testament, they argued that the Earth was at the centre of the cosmos and was surrounded by the firmament, which contained the Sun, Moon, planets and stars. Beyond the firmament were physically located the kingdom of eternal darkness and the place of ascension to Heaven.[75] There were many versions of 'Newtonianism' in this period, and for the Muggletonians 'Newtonian' cosmology was often little more than a shorthand for heliocentrism. Martin did not place the Earth at the centre of the universe but there are, nevertheless, significant parallels between the Muggletonian world system and the Martinian one. Both were plebeian in origin. Both claimed to be accessible because they were founded on common sense rather than esoteric learning. And both emphasized the Word of God rather than human experiment or observation as the true foundation for cosmological knowledge.

A further group of anti-Newtonians to insist on the primacy of Scripture over experiment were the followers of the autodidact John Hutchinson (1674–1737), whose influence persisted into the nineteenth century.[76] Hutchinson was a land steward and surveyor for Charles, the sixth Duke of Somerset. Despite having little formal education, he acquired a scholarly knowledge of Hebrew through private study and also knew some mathematics. He became convinced that the Hebrew language had been corrupted by the relatively recent introduction of the pointing system, and that proper interpretation of the original Hebrew Old Testament would reveal everything needed to understand the universe. Rising to the task himself, and focusing especially on Genesis chapters 1 to 10, Hutchinson developed a system of natural philosophy which centred on three primary physical agents: fire, light and air. These, he argued, were analogous to the three parts of the Holy Trinity: God the Father, God the Son and God the Holy Spirit,[77] and thus his philosophical system confirmed that God was indeed three in one. As Martin was to do almost a century later, Hutchinson personally vilified Newton, insisting that 'as long as Gravity stands, Moses cannot be explain'd'.[78] His alternative, *Moses' Principia*, published from 1724, had relatively little impact when it first appeared, partly because of the contorted, inaccessible style in which it was written. However, a small band of learned disciples set out to spread the word of his philosophy, publishing a collected edition of his works in 1748 which helped bring greater attention to the cause. These disciples were not of plebeian origin, like the Muggletonians, but university-educated gentlemen: as Nigel Aston puts it, 'though it had its fair share of both, Hutchinsonianism was emphatically not a creed for either outsiders or losers'.[79] Hutchinsonians continued to challenge the dominance of the 'Newtonian' experimental philosophy until well into the nineteenth century.[80]

There are many similarities between the Hutchinsonian and Martinian efforts. Martin may have been influenced by the Hutchinsonians when he identified 'Air' as the cause of all things. Both Martin and the Hutchinsonians insisted on the primacy of scriptural exegesis, over observation and experiment, as a philosophical tool, although the Hutchinsonians' emphasis on complex, esoteric, exegetical techniques is in contrast to Martin's unlearned, intuitive approach to Scripture. Martin and Hutchinson both attacked Newton for big-headedly taking personal credit for his philosophical 'discoveries': Hutchinson insisted it was given already by God in the Bible, and Martin claimed that God had inspired him directly to come up with the new system. And both believed their philosophies revealed essential truths about God's role in maintaining the universe.

This latter point is important when it comes to understanding the passion with which Martin communicated his philosophical ideas. It is significant, for example, that his perpetual motion device was powered by a pipe *under the floorboards*. In 1807, when he was exhibiting the device to the likes of Charles Hutton, Martin did not reveal the secret of the pipe to viewers, but left them to wonder at the motion's cause. In 1821, however, describing the device in the *New System*, he is quite explicit about the cause. He admits that, at the time, the cause was concealed and that many people tried to figure it out and could not; yet, in 1821, he certainly did not consider himself as having tricked his viewers. The concealed pipe was not, for Martin, a cunning deception. It was a legitimate invention which was coupled to a highly significant philosophical principle. 'The Perpetual Motion', for Martin, was not *just* a spectacular machine. It was also the principle which kept the universe in motion. 'Perpetual Motion' (of which Air was the cause) kept the planets moving in their orbits; kept the Earth rotating and oscillating; kept living things breathing and multiplying. It was the means by which God motivated his Creation. The Martinian and Hutchinsonian philosophies both implied a reflexive relationship between philosophy and religion. Scripture could reveal truths about the physical universe to the natural philosopher; equally, natural philosophy could reveal truths about religious matters. Hutchinson argued that his philosophy proved the tri-personal nature of God; Martin also used his philosophy as a weapon against Unitarians, as we shall see. Moreover, Martin believed that his philosophy could 'clear up' the Word of God on matters as important as the second coming of Christ and the end of the world.

On a clear frosty night, 29 March 1826, 'a phenomenon' appeared in the sky. Gentlemen living near Martin argued that it was a lunar rainbow, but Martin recognized it as one of the 'signs and wonders before the end of the world' foretold in the Bible. He prayed to God to enlighten him as to the meaning of the sign and it turned out to be concerning his discovery of the new system of natural philosophy upon the principle of perpetual motion. The 'phenomenon' made an arch from east to west and swelled in the centre, which, Martin explained, signified that England was home to true philosophy, arts and science, and a complete revolution of the laws of nature. That the phenomenon was pointed at each end showed that other nations would be very jealous and their proud hearts would be stung. The most significant aspect of the phenomenon, however, was its precise location. 'What is very remarkable', he explained, 'it came over the house that I lived in: I think it was 9 o'clock when the people called me out of my bed to give my opinion on the alarming circumstance.'[81] Like the star of Bethlehem, the 'phenomenon' shone over Martin's house and the people flocked from all around to hear the wonderful news.

Martin's philosophy was closely tied to his religious beliefs. He was a defender of the Anglican Church: he worried that the church was becoming corrupted but, unlike his Methodist brother Jonathan, he insisted that provided certain reforms were made, 'there needs no better Religion than the Church of England'.[82] In practice, despite this stated allegiance to the established Church, Martin was unorthodox in many of his religious beliefs. He styled himself as a prophet, claiming to be in direct communion with God, and he believed that his philosophy had an important role to play in connection with the Millennium, the thousand-year reign of the kingdom of God on Earth before the last judgement.

For example, Martin interpreted the forty-first chapter of the Book of Isaiah verse by verse: each turned out to concern his philosophy. Isaiah 41:21 reads: 'Produce your cause, saith the Lord; bring forth your strong reasons, saith the King of Jacob.' Martin's main bone of contention with the Newtonian philosophy was that it showed only the effects; his own system, by contrast, showed causes – specifically that air was the secondary cause of all things. '[E]ver since I discovered the cause of perpetual motion have I not demanded of the learned men to produce the cause, and they are all ignorant of it?' he wrote in his interpretation. 'You may see clearly that it is the word of God cleared up by my philosophy, that proves Newton and all others who have got the name of philosophers to be impostors and deceivers of mankind.'[83] Isaiah 41:25 reads: 'I have raised up one from the north, and he shall come upon princes as upon mortar, and as a potter treadeth clay.' Reflecting on the geographical location of his home town, Newcastle, Martin concluded triumphantly, 'The Christian philosopher is the man from the north.'[84]

Martin also believed his philosophy to have been prophesied in the Book of Daniel. Daniel was visited in a vision by a man clothed in linen, who spoke to him of a time of trouble that would come before the time of the end. The man said to Daniel: 'But thou, O Daniel, shut up the words, and seal the book, even to the time of the end many shall run to and fro, and knowledge shall be increased' (Daniel 12:4). Daniel 12:4 had long been significant to men of science. On the title page to his *Instauratio Magna* (1620), Francis Bacon depicted a ship of knowledge passing through the pillars of Hercules, and he inscribed in Latin at the foot of the image the quotation 'many shall pass through and knowledge shall be increased'. Bacon believed that this verse showed that the advancement of navigation and exploration and the advancement of philosophy were predestined to coincide in the same – his own – age.[85] From the seventeenth century onwards, Puritan millenarians used Bacon as an authority to lend weight to their convictions that the advancement of knowledge, if sufficiently supported and encouraged, would ultimately bring about a Golden Age.[86] Martin interpreted the verse as Daniel's foretelling the discovery of the Martinian philosophy: 'Now mark this – here is Daniel's prophecy fulfilled, in that part, in the 19th century: "knowledge shall be increased;" that is, the Martinian new system of philosophy shall appear, by the will of God, as foretold by Daniel the prophet.'[87]

The word of God confirmed Martin's divinely ordained status as a Christian philosopher; conversely, the events of Martin's life were useful for clearing up the word of God. Regarding the issue of knowledge being increased, Martin wrote:

> It is a sufficient proof, if we believe God's word, sent forth by his holy prophet, that now is approaching the latter end of time, and the last resurrection. Respecting these things the prophet Daniel and W. M. are perfectly agreed.[88]

Martin explicitly compared himself to the prophet Daniel. Concerning one supernatural vision, he wrote in 1839 that God 'enlightened me to interpret it, the same as he did Daniel, when he caused him to interpret the hand-writing on the wall to the impious King Belshazzar.'[89] The spirit which guided Martin was the same that guided Daniel to interpret the fearful events depicted in John Martin's famous painting.

> The Scriptures state that all people of the world have to belong to one religion before the end of the world. This, Martin argues, requires there to be one philosophy. Now that Martin has discovered what he believes to be the true philosophy, soon all people will belong to one religion, which recognises God as three in one (and does not claim that Jesus was a mere mortal, as some suggest).[90]

Martin believed that once his philosophy was spread all over the globe, the second advent would happen; he therefore took it upon himself to ensure that this was achieved. His performances were not just about getting support for a personally favoured system of philosophy and making a bit of money in the process; they were about bringing on the end of the world. Martin's scriptural interpretations were very unorthodox, and thus contributed to contemporary perceptions of him as an eccentric figure. And yet, at the same time, they drew on a tradition of millenarian prophecy which still attracted followers in this period. While the responses of his opponents indicate that many people did object to his approach, others may have found his combination of scriptural prophecy and natural philosophy more acceptable.

Millenial theology was derived primarily from interpretations of the books of Daniel and Revelation. The practice of correlating the chronologies encoded in these books with historical events in order to arrive at a date for the second coming was common practice for orthodox theologians throughout the eighteenth century, and in the first half of the nineteenth century the events of the millennium were routinely preached in orthodox congregations. However, announcements that the millennium was to be expected very soon were viewed with concern, and individuals who claimed to be in direct communion with God on the issue could cause alarm within the established Church. Nineteenth-century beliefs about the millennium have been categorized in different ways by historians. J. F. C. Harrison makes a distinction between 'millenialism' and 'millenarianism' based not on differences in belief, but on the manner in which such beliefs were established and used. Harrison associates millenialism with the theologically sophisticated traditions of elite eighteenth-century divines. By contrast, he associates millenarianism with a less sophisticated nineteenth-century tradition:

> It is the tone and temper of the popular millenarians, the way in which they used the texts and symbols from Daniel and Revelation, which is distinctive. They were the enthusiasts, the fanatics, the come-outers. Their beliefs were derived from a literal, eclectic interpretation of the prophetic scriptures, and a divine revelation vouchsafed to them directly. A simplicity, often crudity, seemed to mark their mentality, for their reliance on the supernatural enabled them to dispense with many of the limitations imposed by logic and reason.[91]

By Harrison's criteria, Martin was a millenarian. More recently, scholars such as Iain McCalman and Kevin Knox have shown that historical actors did not perceive such a sharp distinction between polite and respectable millennialist scholarship and the unlettered enthusiasm of figures who, like Martin, had their roots in popular culture.[92] Nevertheless, Martin's biblical interpretations can usefully be compared to those of figures whom Harrison identifies as millenarians, figures such as the extremely successful self-styled prophet Richard Brothers.

Brothers was born in Newfoundland in 1757 but lived in England from 1771. In 1790 he began to be enlightened by the Spirit of God and in 1794 he published *A Revealed Knowledge of the Prophecies and Times* in two volumes.[93] Brothers's interpretations of Isaiah, the prophecies of Daniel, the apocalyptic passages in John's gospel and the Book of Revelation were in line with Protestant orthodoxy at the time; he was unorthodox, however, in claiming to have arrived at his interpretations through prophetic visions.[94] Like Martin, Brothers combed the Bible for 'some of the prophecies which mean myself'.[95] Some of Brothers's contemporaries believed him to be a charlatan or madman, yet he had a huge following, which included numbers of distinguished public figures: Members of Parliament, physicians, mathematicians and natural philosophers, such as Peter Woulfe, FRS and recipient of the Society's prestigious Copley Medal.[96] As Harrison and Knox have both shown, it may be difficult for us today to imagine that prophets such as Brothers or Martin could be taken seriously, but this is really 'a measure of our anachronistic blindness'.[97] Prophetic discourse in this period 'was not self-evidently irrational'.[98]

It was not simply the content of Martin's interpretations but also their mode of presentation that was distinctly millenarian. A striking feature of 'The Downfall of the Newtonian System' (see Figure 2.3), for example, was that it was 'Printed verbatim from the original Manuscript':

> Although I am a Cuntry John I must tell unto you all
> By the power of the alwise God I can tell you trouth that shall,
> To the end of the world as shour as the marbel rock
> All the storms of the Divel and Hell cannot it shock,
> Now all the people of Newcastle they may see clear
> Those that interfears with the New System there is nothing
> them for to chear ...[99]

Martin's written English can be better understood within the context of the millenarian prophetic genre. Brothers, for instance, was frequently ridiculed on account of the 'errors' in his writing. However, he explained that initially he wrote in a biblical style using 'thee' and 'thou', but then God ordered him to 'write in the same manner as I speak to you; write as other men do ... according to the custom of the country you live in'. Brothers's learned disciples interpreted the crudity of his writing as evidence that he was a prophet chosen to communicate the word of God to the masses. Nathaniel Halhead, a distinguished oriental scholar and Member of Parliament, argued that precisely because they were 'replete with grammatical faults; destitute alike of harmony and arrangement, and elegance of diction', Brothers's books were 'precisely suited to the comprehension of the most ordinary capacity'.[100]

By contrast to the 'false philosophers' of the day, who pretended to be 'great' and 'learned' men, Martin saw himself as a 'simple individual', able to communicate only thanks to divine inspiration. On the title page to his *Challenge to the Whole Terrestrial Globe*, an eighteen-page pamphlet published in December 1829, he explained:

> You see, gentlemen, I am not a learn'd
> Man; my grammar bad, my spelling the same;
> Yet nevertheless my God doth cause
> My pen to go with a heavenly flame[101]

Martin constantly restated his humble origins and took great pride in the fact that he was self-taught. He placed emphasis on the immediacy of his communications, which were written in a direct, inspired way rather than through clever manipulation of language. This attitude did not apply solely to writing. Of a self-portrait which he used in many of his pamphlets, he wrote:

> I am not a professional engraver,
> Although I completed my portrait in one day;

And that is the truth, and this
Small book, in two, without a lie I say.[102]

The portrait is shown in Figure 2.6. Copied from an engraving by Henry Perlee Parker, its crude outlines, ineffective hatching and irregular hand might, for some, reveal its creator to be a failure at the art of engraving; for Martin, however, that he could complete a portrait or write a book at such great speed even though he was uneducated was evidence that he was assisted by God. Similarly, when in 1826 he followed his brother John in exhibiting his own version of *The Feast of Belshazzar*, painted in oils, he boasted that although he was not a professional artist, 'the Picture was designed and executed in the short Space of One Month'.[103] Martin was, he insisted, an inspired communicator, and as such he did not write expensive esoteric learned books; he wrote short, cheap books that anybody could understand.

Martin's interpretative practices intersected not only with popular millenarian hermeneutics, but also with longstanding traditions of secular prognostication which were still current amongst the working classes in the early nineteenth century. Divination, crystal gazing and astrology, for example, all continued to support thriving trades in this period.[104] Indeed, since the 1790s, Britain had witnessed a surge not only of prophetic literature, but also of periodicals devoted to the occult, the astrological and the marvellous.[105] Astrology had declined in popularity amongst the educated classes since the eighteenth century: in 1828, the *Athenaeum* called almanacs exhibitions of 'palpable imposture, impudent mendacity, vulgar ignorance, and low obscenity', complaining that they been 'utterly uninfluenced by any of the modes of thinking which have marked the emancipation of the present generation from ignorance and credulity'.[106] And yet the complaint itself evinces the continuing popularity of almanacs, which continued to sell in the hundreds of thousands. While Martin did not draw explicitly on astrological theory, his interpretations of celestial phenomena, such as the 'phenomenon' which appeared over his house in 1826, would have resonated with astrological practices.[107] Similarly, his interpretations of dreams would not have seemed especially unusual.

The interpretation of dreams was a common part of folk tradition well into the nineteenth century, as evidenced by the popularity of dream books or 'books of fate' in the period.[108] Dream books contained keys to decode particular components of dreams, which could then be put back together to reveal the dream's meaning. A recognized alternative strategy was to interpret the dream as a whole as symbolic of something else. Martin drew upon established precedents in interpreting his own dreams: the dream of 1806 portending his future recognition by the British Government, for example, he interpreted holistically as an allegory of his whole life. He explicitly compared the dream and its interpretation to

Figure 2.6. In designing his self-portrait, Martin copied an existing portrait by H. P. Parker. His silver medal from the Society for the Encouragement of Arts, Manufactures and Commerce symbolizes his proficiency as an inventor, while the pen and books indicate his achievements as an author. Reproduced by kind permission of the Syndics of Cambridge University Library.

Pharaoh's dream of the seven fat and seven lean kine, interpreted by Joseph as portending seven years of plenty followed by seven years of famine in the land of Egypt. Once again, Martin here portrays himself as a prophet who, rather like Joseph, was inspired by God to reveal the meanings of the signs of the times.

In his study of prophecy in 1790s London, Kevin Knox argues that London was 'a semantic battlefield; winning the right to interpret specific signs, and particularly signs we would classify as scientific, meant winning the right to predict the future', that is, to say what the future *should* be like.[109] He reveals the extraordinary diversity of claims about natural and supernatural phenomena which readers in the metropolis daily encountered, noting: 'Many of the writers and publishers responsible for this array have been either forgotten or conjured up as mountebank lunatics against whose absurdities may be contrasted the contributions of forthright scientific practitioners; but even a brief survey of these writers shows the extent to which they shared questions, representative texts, interpretative techniques and discursive strategies with their esteemed counterparts.'[110] Here I have compared Martin to Richard Brothers, who, despite the strength of his contemporary following, has since tended to be written off as a lunatic. Knox shows, however, that aspects of Brothers's approach to understanding the place of humans in the universe were shared by figures who now hold canonical status in the history of science: figures such as the astronomer and Copley Medal winner William Herschel, who believed the Sun and the planets to be inhabitable worlds; and the theologian and 'father of modern chemistry', Joseph Priestley, whose attachment to phlogiston can be linked to his millenialist beliefs.[111] Knox similarly identifies anti-Newtonian currents within mainstream scientific debate in the period. While members of the scientific establishment did not tend to denounce Newton as vehemently or as personally did Martin, they did question aspects of the Newtonian philosophy. The expelled Cambridge don William Frend, for example, and Baron Francis Maseres, FRS, crusaded against the 'nonsensical unintelligible jargon' of Newtonians – especially the use of negative numbers – into the nineteenth century.

To reiterate, Martin's status as an eccentric figure was not an inevitable consequence of the *inherent implausibility* of his anti-Newtonian, millenarian claims. Rather, I would suggest that it was because his philosophy existed at the margins of plausibility, and because his pronouncements must have resonated with memories of the anti-Newtonians and prophets who had gone before him, that Martin was able so successfully to capture the attention of the people of Newcastle. The unorthodoxy of Martin's philosophical and religious ideas was certainly implicated in his being labelled 'eccentric' yet, as we saw in the last section, his prominent place within Newcastle's scientific culture was also crucially dependent upon how his audiences responded to those ideas. The content of the Martinian philosophy cannot in itself adequately explain the behaviours of Martin's opponents and disciples. It cannot explain the distinctive forms in which they expressed their hostility or approbation. Indeed, it cannot explain why so many of them went to such lengths to respond to the 'eccentric' natural philosopher at all. What did they have to gain?

Science, Carnival and the Politics of Eccentricity

In *Rabelais and His World* (1965), Mikhail Bakhtin describes 'a boundless world of humorous forms and manifestations'.[112] This world, he argues, served the function of opposing the official and serious tone of medieval ecclesiastical and feudal culture: 'completely different, nonofficial, extraecclesiastical and extra political', it constituted 'a second world and a second life outside officialdom'.[113] Bakhtin's 'second world' was built of diverse materials – 'folk festivities of the carnival type, the comic rites and cults, the clowns and fools, giants, dwarfs and jugglers, the vast and manifold literature of parody'[114] – but uniting these was a common thread: all belonged to a medieval culture of folk carnival humour. In nineteenth-century Britain, and in the north-east especially, carnival humour continued to play an important part in people's lives. The *forms* which responses to Martin took were grounded in carnival culture. Moreover, these responses served valuable, if often polemical, *functions* within the local community: beyond simply providing distraction and amusement in idle moments, carnivalesque Martinian performance facilitated, in Bakhtin's words, 'a special type of communication impossible in everyday life'.[115]

The performative activities of Martin's opponents had their roots in popular custom.[116] The behaviour of disapproving audience members at lectures, for example, was redolent of traditional forms of protest such as the 'riding' – a mock procession performed in ridicule of individuals believed to have offended against community norms.[117] This custom got its name from the practice of literally riding the victim (or a proxy) through the streets on a pole, donkey or cart. Ridings were generally accompanied by 'rough music', a cacophonous noise constituted, according to a description by Thomas Hardy, by 'the din of cleavers, tongs, tambourines, kits, crouds, humstrums, serpents, ram's horns, and other historical kinds of music'.[118] Although Martin was not, as far as I know, actually ridden through the streets of Newcastle, the behaviour of those who attended his lectures and demonstrations was robust and clamorous – at times it was considered more appropriate to describe Martin's spectators collectively as a 'rabble' rather than an audience.[119]

The literary forms adopted by Martin's opponents also had precedents within popular culture. Satire featured heavily. In 1828, for example, G. W. Couper published a small book of *Original Poetry*. The last poem, 'An Epistle to Mr W. M—N, the Northumberland philosopher and poet', began:

> O, Willy M—n, Willy M—n, O!
> Thou who all science *never did not know*!
> Thy wisdom, mighty man, is but so-so.

Couper numbered the lines of his poem and appended a barrage of 'Notes, critical and explanatory' to elucidate the text:

> Line 1. O, Willy M—n, Willy—n, O!
> This is a parody on a feeble line in James Thomson's (the author of the Seasons) Tragedy of Sophonisba ...
> Line 2. Never did not know.
> This is in ridicule of our Great Poet's manner of expressing negations. See his Philosophical Song book.

While each line of the poem was intended to ridicule some specific aspect of Martin's writings, the poem as a whole was designed as a parody of 'serious' works of poetry such as Thomson's *Seasons* (1726–30), which, still popular in the 1820s, was published with copious critical notes.[120] The 'Epistle to Mr. W. M—N' was funny because it inverted traditional literary values. By constructing an elaborate critical framework around a deliberately facile composition, it mimicked the activities of Martin's supporters, who professed to attribute great importance to writings which Couper saw as ridiculous and absurd. 'Mr. Wm M—n', Couper explained boldly, 'by the absurdity of all his projects, calculations, and verses; by the contemptible and disgusting egotism with which his works abound, and by the ridicule which he has endeavoured to cast on certain worthy men, has rendered himself a proper subject for Satire'.[121]

Devices used by Martin's opponents can be traced to the early modern period. The practice of satirizing individuals through derisive proclamations supposedly issued by a mock court was popular from at least the seventeenth century.[122] This technique was employed by the 'Newtonian' schoolteacher Henry Atkinson, encountered earlier as the author of 'Ecce Homo', in a second attack on Martin headed 'Imperial Mandate' (Figure 2.7).[123] A mock warrant for Martin's arrest, 'Imperial Mandate' summoned the philosopher to answer before the High Court of Parnassus 'certain Allegations which shall be then and there preferred against him, by our faithful and well-respected Yeomen and Subjects, COMMON SENSE and PLAIN ENGLISH'. Atkinson's broadside was not intended to mislead anybody into believing that Martin was really in trouble with the law; it was intended as a joke, the thrust of which was that Martin's lack of mastery of written English meant he should not be taken seriously as a philosopher. And yet, in its appeals to 'COMMON SENSE' and 'PLAIN ENGLISH', 'Imperial Mandate' also had more radical undertones, invoking actions taken against Thomas Paine, the Newtonian, rationalist, deist, revolutionary and author of *Common Sense* (1776), and his Newcastle-born follower, the radical bookseller and spell-

IMPERIAL

MANDATE.

𝕳𝖊𝖑𝖎𝖈𝖔𝖓 | APOLLO Magnus et Sublimus of the Realms
of Poetry, Fiction, Music, Metaphor, &c. &c.
to 𝖜𝖎𝖙. | IMPERATOR to our principal Inspector of
Mortal Merit, GREETING.

THIS is to command and strictly enjoin you to take, in-
carcerate, and produce, in Salvum Custodium, before
us, on the Volution of our next Parnassian Visitation,
the body of Gullielmus Martin, Nat. Phil. and Poet, to
answer before us to certain Allegations which shall be
then and there preferred against him, by our faithful
and well-respected Yeomen and Subjects, COMMON
SENSE and PLAIN ENGLISH, and to abide our Imperial Sentence if convicted; and for your security and
indemnity, in such Arrestation, this shall be your
sufficient Warrant.

Witness, our truly and well-beloved Cousin and Coun-
sellor, IMPARTIAL INTEGRITY, Lord Chief Investiga-
tor of our High Court of Parnassus in our said Realm,

Dated this 27th. Day of June, 5831.

T. Blagburn, Printer, Old Flesh-Market.

Figure 2.7. In its appeals to 'Common Sense' and 'Plain English, 'Imperial Mandate' has a double significance. It mocks Martin's idiosyncratic use of the English language; more subversively, it hints at the radical politics favoured by his opponents. The date, 5831, is a jibe at literalist scriptural chronologies, according to which the Earth was created in 4004 BC. Reproduced by kind permission of the Syndics of Cambridge University Library.

ing reformer, Thomas Spence, whose early pamphlet on 'The Real Rights of Man' had resulted in his expulsion from Newcastle's Philosophical Society in the 1770s. Here we begin to catch a glimpse of some of the deeper political concerns underlying the activities of Martin's audiences.

Ridings, rough music and mocking rhymes are usually explained as a means by which communities could express symbolically their disapproval of certain behaviours. E. P. Thompson has suggested that during the nineteenth century these customs functioned as 'a licensed way of releasing hostilities which might otherwise have burst beyond any bounds of control'. The comic abuse hurled at Martin can thus be interpreted as controlled expressions of hostility felt towards the philosopher and the ideas he propounded.[124] In the last section, I showed that, despite the seeming absurdity of Martin's philosophy from a modern-day perspective, we cannot assume that he was as easily dismissed in his own time.

Owing to the pervasiveness of prophetic language at the time it was, in Knox's words, 'difficult to exclude the "lunatic visionaries" from conceptions of plausibility', and it is quite possible that there were people in Newcastle who viewed Martin's New System as a plausible alternative to orthodox Newtonian philosophy.[125] Martin's opponents belonged to a sector of Newcastle society who would have been deeply concerned by Martin's popularity, and would have taken the Martinian threat seriously.

All of the opponents whom I have been able to identify were rational dissenters, many of them members of the congregation of William Turner, the dissenting minister and founder of Newcastle's Literary and Philosophical Society. While Turner did not apply the name 'Unitarian' to his congregation, insisting on the freedom and responsibility of each member to form his or her own doctrinal views, Martin's opponents seem mostly to have been Unitarians. (Unitarians made up only around 2 per cent of dissenters in England at this time, but they were concentrated in certain provincial centres, one of which was Newcastle upon Tyne.) Henry Atkinson was a Unitarian, as was Eneas Mackenzie, the radical printer and founder of the Mechanics' Institution, who is named as an opponent in a broadside by the sign-painter Robert May. May himself was evidently a Unitarian too as in a satirical 'Recantation' – May claimed that Martin had asked him to write it down for him because his spelling was so bad – he has Martin uttering the words: 'The Unitarian he is right,/ Consistent is his creed,/ I wish I always had been one –/ Now I am one indeed.'[126] A further, highly influential, opponent was the Unitarian William Mitchell, proprietor of the radical *Tyne Mercury* newspaper and Secretary of the Newcastle Mechanics' Institution. The print trades were especially well represented amongst Martin's opponents: for example, Robert May also names 'Gooseburn', the stationer Robert Gisburne (the misprint was fortuitous for Martin, who henceforth referred to Gisburne as 'the useless old Gander').

Martin's most vociferous opponents thus were high-profile members of Newcastle's rational dissenting community. This community 'exerted a disproportionately large influence on the direction and course of Novocastrian [scientific] institutions'.[127] In the nineteenth century, Nonconformist religion was linked closely to nonconformist politics.[128] Unitarians were so called because they did not believe in a triune God, but they also took a distinctive stance on social reform, education and science. This was grounded in a theory of mind according to which sense perceptions determined character and character determined actions.[129] Based on this philosophy, Unitarians, like other rational dissenters, believed that education of the lower orders could bring about moral improvement and thus lead eventually to social reform. Science education was considered particularly suitable for this purpose, as it encouraged rational thinking and was potentially useful to society. This, in simple terms, was the philosophy behind the foundation of Mechanics' Institutes and the proliferation of affordable, 'improving' literature, such as that produced by Henry Brougham's Society for the Diffusion of Useful Knowledge (which William Turner energetically supported).[130] In Newcastle, Turner's chapel at Hanover Square was a 'visible symbol of the marriage between liberal politics, liberal religion and useful knowledge'.[131] To this community, Martin symbolized the opposite: unreasonable attachment to orthodox religion and the production and public dissemination of (mostly) useless inventions and irrational philosophy. It thus comes as little surprise that Turner, Mackenzie, Mitchell and their associates did not allow Martin to lecture within the scientific and educational institutions over which they had control. Equally, it is understandable that the Newtonian, Unitarian schoolteacher Atkinson should have gone to some trouble to expose Martin as an 'ignoramus' and madman. That Martin's opponents disagreed with his philosophical and theological claims can begin to explain their remonstrances. And yet this cannot be the whole story, because, for a start, it leaves entirely unexplained the activities of Martin's 'disciples'.

While Martin's opponents were mainly rational dissenters, his 'disciples' were mainly Tory Anglicans. His most influential supporter was Edward Walker, proprietor of the Tory *Newcastle Courant* – probably one of the 'wags of the press' referred to in the *Durham County Advertiser* report on the Northumberland Eagle Mail demonstration. Originally from York, Walker moved to Newcastle and purchased the *Newcastle Courant* in 1795; under his ownership it became highly profitable, with the highest advertising rates and largest circulation of any Newcastle newspaper.[132] Walker printed and probably subsidized Martin's lengthy advertisements on the front page of his newspaper almost on a weekly basis during 1827 and 1828, suggesting that he found in Newcastle's Tory reading public a favourable and receptive audience for Martin's performances. In the last section I suggested that some Newcastle inhabitants may have taken Martin seriously, considering his New System to represent a realistic challenge to the Newtonian philosophy. Some readers of the *Newcastle Courant* may have felt this way. However, a closer look at the published responses of Martin's 'disciples' reveals that at least some of those disciples did not take Martin seriously at all. They did not, it turns out, 'support' him because they necessarily *agreed* with him on philosophical and religious matters, so how can their activities be explained?

Throughout his career, Martin stressed his Northumberland origins. He claimed to be patronized by the Duke of Northumberland (in reality he requested patronage and interpreted the Duke's failure to reply as confirmation of his support); he worked 'Northumberland' into the titles of his publications and the names of his inventions – the Northumberland Eagle Mail, for example; and in connection with his poetry he styled himself 'the Northumberland Bard'.[133] Martin's disciples highlighted the philosopher's regional affiliations in their writings. Some wrote in the Northumbrian dialect.[134] Others explicitly contrasted Northumbrian Martinian philosophy with the supposedly inferior philosophies of the south. In celebrating Martin as a peculiarly *local* eccentric character, Martin's Tory disciples were commenting more generally on the distinctive character of Tyneside, as something to be preserved against what they perceived as the centralizing, homogenizing threat of the political radicals and reformers (who in this context were Martin's opponents).

In the summer of 1828, for example, at the peak of Martin's lecturing career, a Mr Walker of London arrived in Newcastle to deliver a course of astronomical lectures at the Newcastle Freemasons' Lodge. Strangely, at precisely the same time, a Mr Lloyd, also of London, brought his 'grand transparent orrery' to be displayed at the Newcastle Theatre Royal; advertisements promised that astronomical knowledge would be divulged 'in the same elementary and scientific manner as given in the Universities of Oxford and Cambridge'.[135] No sooner had the advertisements appeared than Martin and his supporters were up in arms. The first attack appeared on 11 June: 'Martin Victorious!', an anonymous broadside lampooning the London astronomers, printed by Edward Walker.[136] The broadside contained two poems. The first, to be sung to the tune of 'Scots Cam' o'er the Border', reassured the people of Northumberland that Martin would defend the region against the impending invasion from the south: 'Have you seen two London astronomers,/ With their dashing wigs out of order?/ MARTIN has clipp'd their soaring wings,/ And they'll ne'er fly over Northumberland border.' The second, composed by 'T. G., a Disciple of Mr M', related a dialogue between the two London astronomers, supposed to have taken place during their preparation for lecturing in Newcastle. The premise for the poem was the paradox of why two astronomers should come all the way from London to lecture on astronomy at the same time. The solution offered by T. G. was that these '*Noodles*' were desperate to put down the New System of natural philosophy, but each feared he would be unable to take on Martin single-handedly. Together they felt they might just have a fighting chance. A couple of days later, Martin reacted in an announcement of his own. Explaining that the London astronomers' orreries, spectacular as they may be, were founded on false principles, he warned discerning readers to avoid them all costs. (Later in the month he reiterated his reservations to Lloyd in person when he attended one of his lectures at the theatre.)[137] In attacking the London astronomers, he appealed to the regional pride of the people of Northumberland. 'If the Gentlemen of London, Cambridge, and Oxford', he wrote, 'suffer themselves to be imposed upon by those silly Things, which are no better than Punch and Toby Shows, the Exhibiters of them must not think that they will be permitted to do the same in Northumberland, – for Northumberland has become the School of true Philosophy for the whole World!'[138]

Martin was not, as it happens, the only local eccentric character of his day. Figure 2.8 is an engraving of a painting called *Eccentrics and Well Known Characters of Newcastle*. It was painted around 1818 by the artist Henry Perlee Parker who, having recently arrived in the region from Devon, made his name by depicting local characters – he did Martin's portrait at a later date. The original painting, *Eccentrics and Local Characters*, was bought by the Tory MP for Northumberland, who became a patron of Parker.[139] Copies of the engraving could be found in 'numerous households in Newcastle and the vicinity', and were popular with older residents into the 1890s.[140] The characters portrayed were remarkable for a variety of qualities including blindness, feebleness, penury, musical ability, partiality to drink, vagrancy and dietetic experimentation; a few are mentioned by name in Martin's writings and those of his respondents. Martin shared certain characteristics with some of them: 'Captain' Benjamin Starkey (*c.* 1757–1822), an inmate of Freeman's Hospital and then the poorhouse in Newcastle, was given the title of Captain on account of his vanity and pretension to grandeur, something Martin was also accused of; James Brown (d. 1823) was a follower of the millenarian prophetess Joanna Southcott. But while, in a sense, Martin had rivals, his involvement with natural philosophy, millenarianism, pamphleteering and popular lecturing, coupled to his overt loyalty to Church and State, made him uniquely useful as a symbolic figure through which Newcastle's Tory Anglicans could assert their views, and challenge those of their opponents, on highly contentious political issues such as freedom of expression, religious toleration and educational reform.

On 3 April 1827, the day before Martin's first lecture at the Dun Cow, a letter addressed to the editor appeared in the *Tyne Mercury*. It was signed 'Fair Play', and its manifest purpose was to chastise the editor, William Mitchell, for excluding a notice of the forthcoming lecture in a previous issue of his newspaper: 'It very ill accords, let me tell you, Sir, with your boasted liberality, and the principles of a free press, to refuse admission to any remarks made by a sensible man on a subject so highly important.'[141] Fair Play was evidently a self-styled Martinian 'disciple'. If, as Fair Play suggested, Mitchell received a notice of Martin's forthcoming lecture, it is possible that he excluded it on principle. While he printed occasional broadsides and pamphlets for Martin, this work was undertaken in a commercial context, in which it was understood that the product was not representative of the printer's own views. It was common for newspapers to take a similar stance towards advertising and announcements in this period; however, the *Tyne Mercury* took the line that editorial opinion was to be reflected in the advertising pages as well as in the editorial sections – Mitchell's father, John Mitchell, had taken the first step in 1812, pledging that his newspaper would no longer carry those 'disgusting and sickening Advertisements of the poisonous trash known by the name of Quack Medicines'.[142] In his letter,

Figure 2.8. *Eccentrics and Well Known Characters of Newcastle* (*c*. 1818), by H. P. Parker. Billy Scott (sixth from the left) and Doodimdaddim (far right) are mentioned in Figure 2.5 as being marginally less crazed than the Newtonian author of 'Ecce Homo' and 'Imperial Mandate'. Reproduced by kind permission of the Laing Art Gallery, Tyne & Wear Museums.

Fair Play turned this situation around. Affecting outrage that such an important event had failed to be accorded proper notice, he was able to satirize the *Tyne Mercury*'s liberal politics and thus accuse Mitchell of hypocrisy for attempting to silence an individual (Martin) whose views were opposed to his own. Though apparently concerned only with the Martinian philosophy, the letter was meant satirically to undermine the editor of the *Tyne Mercury*, his radical newspaper and the liberal politics of its readers.

Getting into his stride, Fair Play went on to suggest an alternative explanation for the non-appearance of the notice of Martin's lecture: 'I am afraid you feel a little sore at your not having the philosopher's advertisement for your paper, but this could not be, for Mr Martin being an *out-and-outer* in orthodoxy himself, could only confer *this honour* upon those in whose faith there is no flaw; and, let me tell you, Sir, he strongly suspects there is a flaw in yours.'[143] Mitchell was a Unitarian and a member of William Turner's dissenting congregation, in which members were encouraged to form their own doctrinal views based upon careful reading of the Scriptures. Martin was nominally a supporter of the Church of England, which took a more prescriptive line on such matters, and evidently Fair Play was too. In his letter about Martin's lecture, Fair Play took the opportunity to satirize the religious views of Martin's opponents. Martin claimed to be violently opposed to religious unorthodoxy, the obvious irony being that many of his ideas were extraordinarily unorthodox, verging at times upon the heretical. Fair Play used this irony to its full potential: 'Is it not, then, a disgrace to the age in which we live', he wrote, 'that a mind of such amazing calibre as that man must possess, who has discovered that the Deity resides in a palace in the sun, should be so persecuted – should have the privileges of the press denied him?'[144] Through a process of symbolic inversion, Fair Play raised up Martin – a self-styled prophet who claimed that God lived in a palace in the Sun – as a paragon of orthodoxy. On the surface, Fair Play claimed to 'support' Martin, but at a deeper level he was setting him up as a warning. Allowing the uninstructed multitude to form their own opinions on doctrinal matters was, in Fair Play's opinion, a recipe for disaster, and the extreme unorthodoxy of Martin's views was evidence for this.

After several more allegations, one threat and much praise of the New System, Fair Play's letter concluded with a postscript: 'I have been told that Mr Martin has been named as one of the professors in the new London University. The founders could not have given a stronger proof of their judgement and discrimination.'[145] The London University had recently been founded by reformers as an alternative to Oxford and Cambridge for young men from dissenting backgrounds. Placing a strong emphasis on scientific training and wider access to education, it had close ties with other reformist institutions, such as the Mechanics' Institutes, with which Mitchell and other of Martin's reformist opponents were actively involved. The idea of Martin being made professor at the London University, or indeed anywhere, would naturally have been preposterous. Once again, however, through a process of symbolic inversion – through promoting the self-taught eccentric philosopher temporarily to professor – Fair Play was able to ridicule those whose political opinions differed from his own, deriding the 'judgement and discrimination' of the University's founders and their supporters. In a postscript to a second letter, sent the following week, Fair Play wrote, 'I have heard ... that the Society for the Diffusion of useful Knowledge mean to apply to Mr Martin for a *Treatise on Astronomy*.'[146] Once again, Fair Play used Martin as a means of satirizing a reformist educational institution, this time the SDUK, which produced affordable 'improving' books aimed at working-class readers. Efforts to open up learning to the masses and to improve the lower orders through scientific education were, in Fair Play's opinion, futile, even dangerous. Plebeian access to scientific knowledge would lead to philosophical nonsense, false confidence and – as Martin's lectures demonstrated – uproarious, threatening behaviours.

The pseudonym 'Fair Play' has significant connotations. It implies that the letter is written in the spirit of play. It thus sets up the context for any further exchange as a game – a set-apart world in which rules different to those of ordinary life apply. Within this game, all conduct, provided it adheres to the game's rules, is to be considered 'fair play', is to be considered upright, even if such conduct might be considered inappropriate in real life. These letters to the *Tyne Mercury* were an opening gambit. They marked the beginning of an extended period of intense investment in Martin and his philosophy, from both sides of the political divide. From this point on Martin's disciples, and to some extent his opponents, engaged with Martin playfully, through a rich mix of carnivalesque cultural performances.

Martin's disciples, like his opponents, drew on established conventions in formulating their literary and performative responses to him. Parody was a prime means through which they satirized their opponents. Simply the act of styling oneself 'A Disciple' was a sort of parody: of the disciples of Jesus, of the devoted followers of the anti-Newtonian Hutchinson, of the millenarian prophet Brothers, or of the radical Thomas Spence (whose disciples continued to champion the cause of radical reform after his death in 1814).[147] In their literary effusions, disciples aped genres of eulogic poetry traditionally reserved for celebrated rulers, statesmen and other public figures. For example, in Chapter 1, I showed that celebratory poems of the seventeenth and eighteenth centuries often described heroic figures as being eccentric *like comets*; Martin's disciples employed this figure in their own declamations:

> And when your body low may lie,
> Deep hid from every mortal eye,
> Among the stars you're sure to fly
> A comet ! Mr. Martin

They also parodied ecclesiastical genres such as hymns and other religious orations, a tactic adopted by Spence's followers.[148]

Scattering their poems with clever puns and silly metaphors, Martin's disciples extended their praise to ridiculous extremes:

> GREAT MARTIN ! thy works are *sublime*,
> I've read them with *great satisfaction;*
> I know they'll stand *proof against time*,
> Now Sir Isaac has lost his attraction.[149]

Elevating the self-taught philosopher to the level of a sage, they named him 'Great Martin', 'Professor Martin' and 'Champion of Truth'; they even attributed to him divine qualities, calling him a 'god-like man', a man whose 'genius is divine'.[150] Traditionally esteemed figures, by contrast, were debased. Newton,

Boyle, Locke, Ptolemy, Leibnitz, Descartes, Milton, Shakespeare, Swift, Dryden, Pope: the whole canon of scientific, philosophical and literary heroes was ejected in Martin's favour.[151] Newton, in particular, was subjected to silly, childish forms of abuse: 'Sir Isaac Newton was a cheat', wrote one disciple; he 'told such fibs' agreed another.[152] Several noted that the *Principia* had been rendered waste paper, and amused themselves with suggesting alternative uses for the redundant leaves of the once-sacred text.[153]

The activities of Martin's disciples, and some of his opponents, inverted traditional hierarchies. In thinking about how these activities can be better understood, I have been influenced by anthropological and literary-critical writings on rituals of inversion and carnival. In the 1950s the anthropologist Max Gluckman explored the role of inversion in 'rituals of rebellion' in south-east Africa – rituals in which the subordinated are temporarily given licence to assert their dominance: women over men; subjects over king. Gluckman argued that rituals of rebellion occur only within established and unchallenged social orders. By openly expressing tensions between particular authorities and individuals, they emphasize the social cohesion within which such conflicts exist: the dramatic action of ambivalent social relations is meant ultimately to achieve unity and prosperity.[154] A similarly conservative function has been assigned to ritual inversion by Victor Turner (at least in his early writings), whose research has addressed the role of such inversions in 'cultural performances' such as carnival. '[I]n cultural performances', Turner writes, 'we are concerned ... with a topsy-turvy, inverted, to some extent sacred (in the sense of "set apart," hedged around with taboo and mystery) domain of human action.'[155] Like Gluckman, Turner sees ritual inversion as reaffirming the hierarchical principle: 'By making the low mimic (often to the point of caricature) the behaviour of the high, and by restraining the initiatives of the proud, they underline the reasonableness of everyday culturally predictable behaviour between the various estates of society.'[156]

The topsy-turvy world of Martinian philosophy was a world set apart from everyday Northumberland life – a world in which social and intellectual hierarchies were inverted, traditional epistemologies were swept away, and ordinary rules of conduct were laid temporarily aside. In Martin's world, dogs barked sense, school teachers were corrected by idiots, silly little men stood, literally, on the crowns of their heads, and local worthies jeered and jostled with the common people, hurling preposterous praise or vulgar abuse at the intractable philosopher. In so many ways, this represented the very opposite of the calm, rational audiences envisaged by the proponents of Useful Knowledge. Martin's disciples used carnival to highly conservative ends. By encouraging, contributing to and drawing out the carnivalesque aspects of Martinian performance, they were able to set him up as a demonstration of what they felt would be the inevitable, unreasonable consequence of the opposition's ideals of freedom of expression, religious toleration and useful knowledge.

Though Fair Play's choice of pseudonym suggested that he was a good sport, this does not mean that he and the others who responded to Martin were necessarily engaging in harmless fun. While anthropologists such as Gluckman and Turner saw carnivalesque inversion as functioning ultimately to preserve existing social orders through the controlled release of social tensions, other theorists, including Bakhtin and Natalie Zemon Davies, have assigned to it more subversive, rebellious and antagonistic capabilities.[157] Stallybrass and White point out that it 'actually makes little sense to fight out the issue of whether or not carnivals are *intrinsically* radical or conservative, for to do so automatically involves the false essentializing of carnivalesque transgression'; they conclude that while carnival may not always have politically transformative effects, 'given the presence of sharpened political antagonism' it may often act as a '*site of actual and symbolic struggle*'.[158] Fair Play's ambivalent combination of praise and abuse – a combination Bakhtin identified as being central to carnivalesque performance – may have made his letter acceptable, printable, but this did not compromise the hostility of his attack upon Mitchell and his religious and political threats, and it is ultimately unclear whether, overall, carnivalesque Martinian performance served to ease or exacerbate relations between the different parties.

No matter how much analysis he is subjected to, William Martin remains a deeply ambiguous figure. By way of illustration, it is worth briefly comparing his contemporary reputation to those of his brothers John and Jonathan. John Martin, as we have seen, was the most popular historical painter of his day. He accumulated a personal fortune by exhibiting and selling his works and, while he was not formally acknowledged by the Royal Academy, he unambiguously belonged to the cultural elite. He knew intimately many members of the London intelligentsia and, despite his reformist leanings, engaged socially with the aristocracy. Although some, like Samuel Carter Hall, fancied that his genius was not so far removed from the eccentricity and madness displayed by other members of his family, it cannot be doubted that John must have appeared to most observers as a social and professional 'success'.[159] The same cannot be said of Jonathan Martin, a committed lunatic and convicted criminal. Nor can it easily be said of William Martin, although his career is rather more difficult to assess. There were those who accused Martin of being 'cracked', but he was never tried for lunacy or any criminal offence. Conversely, there were those who celebrated him as the champion of true philosophy, but the evidence suggests many of his disciples were, at some level, disingenuous in their praise. Martin's symbolically privileged position within Newcastle as an eccentric natural philosopher depended precisely on this deep ambiguity.

In a devastating critique of Martin's *New System of Natural Philosophy*, a reviewer for the *Newcastle Magazine*, probably the magazine's editor William Mitchell, had observed in 1822:

> Mr Coleridge somewhere says, 'until you understand a writer's ignorance, presume yourself ignorant of his understanding'. But Mr Coleridge forgot to make it a condition in this strange observation, that there might be ignorance which had nothing in it that could ever be understood. It would puzzle even Mr Coleridge's subtlety to find out how one could understand what was not understandable.[160]

Martin's *New System* may have been 'original' and 'amusing' but overwhelmingly, to Mitchell, it just did not make sense.[161] He observed wryly that Martin had prefigured this eventuality, allotting to one of his chapters the 'sage and necessary title, "What is this all about?"' In this chapter Martin had dwelt upon the future reception of the work, envisaging 'honest but untaught' readers grappling with the wonders of the Martinian philosophy just as they might grapple with the 'wonders of the universe' itself. 'When Mr Martin's volume is read in the twenty-first century', Mitchell speculated, 'it will be a question whether he intended to raise his own works by this comparison, or to depreciate the universe; and the wise of that day will doubtless exclaim that, though not quite so extensive as the universe, his works are equally inexplicable'.[162]

Martin's works are extensive to be sure but, needless to say, I do not want to have to conclude that they are inexplicable. The purpose of this chapter has been precisely to explain Martin's philosophical activities – motivationally, in terms of family influences and his personal convictions; historically, in terms of the literary and performative genres which he and his respondents drew upon; and functionally, in terms of the carnivalesque roles which he and his philosophy played within Newcastle society. In fairness to Mitchell, however, I do not think these kinds of explanation were quite what he had in mind. In his review, Mitchell's criticisms were directed very much at the text of the work: at its argumentative structure, which he found unsound, at its language, which he found obscure, and at its composition, which offended against his preconceptions of what a philosophical treatise ought to be like – 'the publication of poems at the end of a system of philosophy', he wrote, exasperatedly, 'is something as original as the system itself'.[163] Literary eccentricity is the focus of the next chapter, which examines in depth the production and reception of two 'eccentric' books by the fossil collector Thomas Hawkins.

3 'BEYOND THE PALE OF ORDINARY CRITICISM': LITERARY ECCENTRICITY AND THE FOSSIL BOOKS OF THOMAS HAWKINS

'What a title is here!' exclaimed an anonymous reviewer of fossil collector Thomas Hawkins's *Book of The Great Sea-Dragons, Ichthyosauri and Plesiosauri,* גדלים חביבם, *Gedolim Taninim, of Moses: Extinct Monsters of the Ancient Earth* (1840).[1] 'And what a book does that title usher into notice! Unquestionably the most strange and extraordinary, in all points of view, that we have been called upon to examine and characterize during the not brief course of our critical career.'[2] *Sea-Dragons* was Hawkins's second book about the fossil skeletons in his renowned collection. His first, *Memoirs of Ichthyosauri and Plesiosauri, Extinct Monsters of the Ancient Earth*, appeared in 1834.[3] 'This is an extraordinary book', was one critic's verdict; 'one of the most singular productions to issue from the modern press', wrote another.[4] Some readers loved Hawkins's books for their originality, humour and 'Quixotism', while others criticized them for their extravagance, obscurity and 'exceeding *bizarreries*'.[5] Some laughed at 'the somewhat eccentric style in which [they were] written and put together',[6] while yet others quite simply did not know where to start. Until very recently, commentators have tended to take the perceived eccentricity of Hawkins's fossil books as a reason for dismissing them.[7] In *Finders, Keepers: Eight Collectors*, for example, Rosamond Wolff Purcell and Stephen Jay Gould describe Hawkins as 'eccentric and demented'. 'The two monographs of 1834 and 1840 display a mind in disaggregation. The 1834 work ... is florid enough, but still tractable. The 1840 text is all but unreadable.'[8] Contemporary readers explicitly labelled the books 'eccentric' too, but they considered them worthy of further criticism, even if this rarely turned out to be a straightforward task.

This chapter takes an alternative approach to *Memoirs* and *Sea-Dragons*, analysing them precisely as specimens of 'eccentric' scientific literature. The last chapter looked at how Martin used varieties of *public performance* to construct and manage his eccentric identity; this chapter focuses on eccentricity as mediated through *writing*. Part of the story will thus be about how Hawkins constructed an 'eccentric', authorial persona through his writings, and how this authorial persona was understood by his contemporaries. The focus, however, is on the books themselves as eccentric objects.

In Chapter 1, I showed that it was not just people who could be described as 'eccentric' in the nineteenth century. Here, I suggest that Hawkins's books were labelled 'eccentric' because, like the misers, hermits, hermaphrodites, giants, free-spirited women and unfashionably dressed gentlemen who graced the pages of nineteenth-century eccentric biography, his books seemed to challenge classificatory boundaries. Specifically, they seemed physically and textually to undermine the generic boundaries which conventionally defined the different kinds of literary work.[9] Of the three case-studies in this book, then, this is the one most explicitly concerned with the relationship between boundaries and eccentricity. Hawkins's case provides a vivid illustration of how 'eccentric', boundary phenomena were interpreted, discussed and debated in the period. It also exemplifies how studying 'eccentric' phenomena can help to illuminate moments at which new boundaries were laid down and new categories defined. In this case, it illuminates the processes by which new disciplinary and generic boundaries were laid down in a crucial period in the history of geological science.

The chapter begins by locating Hawkins within nineteenth-century cultures of fossil collecting and geology. The remainder of the chapter focuses on rhetorical and reception analysis of his fossil books. The second section applies techniques of genre analysis to *Memoirs*: I take genre to be defined by a set of conventions and codes, shared by implicit agreement between writer and reader, which make possible both the creation of the work and its interpretation.[10] Drawing on the ideas of the literary theorist Gérard Genette, and comparing Hawkins's books to other geological works of the time, I argue that the 'paratext' of *Memoirs*, and many of its stylistic features, gave it the appearance of an anatomical monograph.[11] As I show in the third section, however, Hawkins's books were generic hybrids. His sporadic inclusion of pastoralized discovery accounts, for example, and his unconventional use of illustrations challenged readers' initial expectations, frustrated their attempts at interpretation, and led them to characterize *Memoirs* and *Sea-Dragons* as 'eccentric' scientific books. The final section shows how further interpretative difficulties were presented by Hawkins's 'enthusiastic', visionary reconciliations of palaeontological science with biblical accounts of the early history of the Earth. These passages were considered by many readers to be inappropriate to an anatomical monograph; to make matters worse, they were written in a self-consciously obscure style: 'Enthusiasm seems as extinct as the Sea-dragons which here inspire it: their strange eloquent Remains bespeak a Chord in our breast, which vibrates only to the Master Touch ...'.[12]

In terms of source material, one advantage of Hawkins's publications over Martin's is that they were fairly widely reviewed in the periodical press; treated cautiously, these reviews can reveal a great deal about what contemporaries thought about them. Tellingly, one of the most frequent complaints of professional critics is that they found themselves frustrated by the task of criticism itself. They struggled to convey the overall sense of the books, they had difficulty identifying comparable works, and they were unable to judge them in terms of success or failure. The reasons behind these complaints are explored in this chapter. In the *Athenaeum*, the young surgeon Percival Lord opened his anonymous, scathing review of *Memoirs* by quoting its opening lines. 'Every generation of man is born to stare at something, which, as long as it eludes their understanding, is a very African Fetishe to the many and a Gordian knot to the few.'[13] While Hawkins meant the analogy to apply to geology, Lord wittily turned it on the author himself: 'though he may comprehend geology, we cannot comprehend him; he is something which 'eludes the understanding'.[14] Through exploring Hawkins's literary practices and contemporary readers' efforts at interpreting them, this chapter provides a framework for better 'understanding' eccentric literary works.

Collecting Sea-Dragons

Geology was a new and exciting science in the 1830s, and oryctology – the study of fossils – was one of its most fashionable branches. As a collector and author, Thomas Hawkins was able to take advantage of the manifold possibilities this new science offered. Hawkins was born in 1810 near Glastonbury in Somerset, England. His father was a farmer and cattle dealer; his mother died when he was just a few months old. Hawkins was intended for the medical profession and studied surgery for a time at Guy's Hospital. When he was just twenty, however, his father died leaving him a modest inheritance; he immediately abandoned his medical training and wholeheartedly applied his inheritance to his favourite pursuit, the accumulation of fossil organic remains.[15] Hawkins later claimed he had always had a passion for collecting. Initially he collected coins, pottery and worm-eaten books but he soon discovered fossils to be more fashionable, more philosophical, and sometimes even more expensive.[16] By his twenty-fourth year he had, according to the geologist and surgeon Gideon Mantell, spent nearly four thousand pounds on his hobby.[17] It is likely that Hawkins saw fossil collecting as a means of advancing his social position. Having been born into a family of middling rank he had much greater social ambitions, and he hoped that by soliciting recognition for his collecting endeavours he would be able to mingle with the elite of gentlemanly geology.

Since the turn of the nineteenth century the Earth had yielded to its investigators natural wonders of unparalleled curiosity. Before the discovery of the dinosaurs, the most spectacular fossil skeletons to come to light were the ichthyosaurs and plesiosaurs, both types of gigantic marine reptile. The first ichthyosaur to come to the attention of gentlemanly geologists in London was found by the young collector Mary Anning, probably with the help of her brother Joseph, around 1811. Anning, who later took over the family fossil collecting business in Lyme Regis, is sometimes credited with finding the first ichthyosaur, though this is slightly misleading: in fact, specimens later identified as ichthyosaurs had been discovered in the eighteenth century, but at the time they were believed to be crocodiles.[18] Nevertheless, her find caused great excitement and some confusion. The famous French comparative anatomist Georges Cuvier captured the sense of bafflement which this animal produced: it had 'the snout of a dolphin, the teeth of a crocodile, the head and sternum of a lizard, the paddles of cetacea [the order including whales, dolphins and porpoises], but numbering four, and finally the vertebrae of fish'.[19] Anning's specimen was technically described in 1814 by the surgeon Sir Everard Home, who claimed it was 'more nearly allied with fishes than any of the other classes of animals'.[20] In 1818 it was sold to the British Museum, where Charles König, Keeper of Natural History, named the genus *Ichthyosaurus*, meaning 'fish reptile'. In 1821, however, the English palaeontologists Henry De la Beche and William Conybeare delivered a paper to the prestigious Geological Society of London arguing convincingly that the ichthyosaur was indeed similar to both fishes and lizards, but was actually far closer to the latter, superficially resembling fish only because its skeleton was adapted for the organism to live in water.[21] An illustration of a fossilized ichthyosaur skeleton from Hawkins's collection is shown in Figure 3.1.

In the same paper of 1821, De la Beche and Conybeare described another remarkable new animal, the plesiosaur. Figure 3.2 is a photograph of a plesiosaur specimen from Hawkins's collection. De la Beche and Conybeare named the genus *Plesiosaurus*, meaning 'near reptile', because they thought it closer to modern reptiles than *Ichthyosaurus*.[22] In the years that followed, yet further exciting discoveries were made. The Oxford geologist and cleric William Buck-

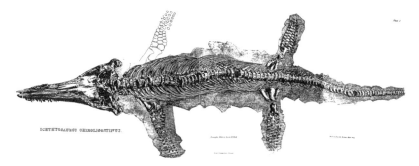

Figure 3.1. Illustration of an ichthyosaur from Hawkins's collection. *Memoirs of Ichthyosauri and Plesiosauri*, plate 3. Reproduced by kind permission of the Syndics of Cambridge University Library.

Figure 3.2. An immature plesiosaur from Hawkins's collection, now on display at the Natural History Museum, London. This particular species was later named *Plesiosaurus Hawkinsii* in recognition of Hawkins's contribution to palaeontology. © The Natural History Museum, London.

land announced the megalosaur, the first dinosaur, to the Geological Society in February 1824. The following year, Gideon Mantell announced the iguanodon to the Royal Society.[23] News of these discoveries of ancient, extinct animals soon spread beyond the elite circles of gentlemanly geology and captured the public imagination. Described in newspapers, magazines and popular books, they inspired enthusiasts from all walks of life to seek out bigger, more complete, or perhaps even entirely novel specimens.

Many of Hawkins's specimens came from the limestone quarries at Street and Walton, not far from his home, in which strata of the Lower Lias from the early Jurassic period were exposed. Hawkins amassed a gigantic collection of specimens of the flora and fauna of this formation. 'I have gathered every thing indiscriminately', he once told Buckland, who would become his mentor, 'until my drawers groan with shells, sauri, plants.'[24] In order to build up his collection, Hawkins established relationships with people who could supply him with specimens. He visited the local quarries regularly and acquainted himself with some of the quarrymen, who soon learned that good money could be made from selling fossils they found in the course of their work. A favourite was George Moon, a labourer at the quarry of Mr Somers in Walton. According to Hawkins, he 'possessed more shrewdness than the generality of his class'.[25] Sometimes Moon, or one of his fellow workers, would show Hawkins an enticing fragment and invite him to inspect the site. If he liked what he saw he would pay a reward for the find, and hire a team to extract the specimen. Hawkins also drew on the expertise of commercial fossil dealers, including Jonas Wishcombe of Charmouth[26] and

Mary Anning. One day in July 1832, for instance, Anning extracted part of the head of an ichthyosaur from a limestone cliff near Lyme; the rest of the animal, she presumed, was buried deep in the cliff and was thus inaccessible. It probably would have remained there had not Hawkins happened to arrive at Lyme the same day. When Anning showed him the skull fragment he immediately set about obtaining the permission and the workforce needed to remove 'as much of the cliff as was necessary' to extract the entire skeleton.[27]

Hawkins was proud of the extreme lengths he was willing to go to in order to obtain desirable specimens. When De la Beche remarked, 'you carry away whole quarries!', he took this as the greatest of compliments.[28] He frequently boasted about the inconveniences – practical, financial and psychological – that he suffered for the sake of his collection, and even compared himself to Cheops, the despotic Egyptian king who forcefully channelled the energies of his entire kingdom towards the construction of the Great Pyramid at Gizeh: 'I went on, gathering one rarity after another, as a second Cheops with a million slaves at his imperial beck might', he wrote proudly.[29] Just as Cheops sent out expeditions far and wide to collect stone and precious minerals for his beloved pyramid, Hawkins saw himself as commanding an almost imperial network of collectors, labourers and artisans, all working towards the completion of his own monumental collection.

Hawkins prepared many of his specimens himself. For each fragment he first removed parts of the matrix – the surrounding stone – from the fossil bones to create a level slab; sometimes he strengthened the bones and remaining matrix with acacia gum. He then placed the slabs together to reassemble the overall form. He generally wall-mounted his specimens, arranging the bones and replica parts in wooden frames and then filling in the gaps with plaster. Finally he painted a slip or varnish over the whole to give a seamless, even-coloured finish.[30] Sometimes he paid other people to assist him. The giant ichthyosaur from Lyme, for instance, was a particularly tricky job. It was enormous – the bones and surrounding matrix weighed a ton – and it was broken into nearly six hundred pieces, many of which were so fragile that it was dangerous to touch them. Hawkins did the initial preparatory work himself, unpacking the fragments, strengthening them with acacia gum, and leaving them to dry in the heat of the July sun. He entrusted the reconstitution of the fragments to an Italian sculptor, who fixed them in three thousand pounds of sulphate of lime, in a case that weighed half a ton.

It was common practice amongst commercial fossil dealers to fill in gaps with bones from other specimens, or with plaster replicas, and Hawkins did the same. The propriety of this practice was contested within scientific circles. Anning, for instance, wrote that Hawkins was 'such an enthusiast that he makes things as he imagines they ought to be; and not as they are really found'. Gideon Mantell wrote 'I think it is objectionable when art is allowed to interfere so far', but conceded that Hawkins 'does not do this without authority'.[31] Nevertheless, all agreed that his collection of ichthyosaurs and plesiosaurs was one of the finest in the world and, to this day, his specimens continue to impress visitors to the Natural History Museum, London, where many are on display.[32] Many of Hawkins's contemporaries, including some very eminent men of science, went to see the collection for themselves. In 1831, for instance, Hawkins displayed part of his collection at The Strand, London. Buckland came to see it, and was so impressed that he offered to introduce its twenty-year-old owner to the Geological Society – Hawkins later recognized this as the moment at which his oryctological fate was sealed.[33] Other visitors of note included Gideon Mantell, who breakfasted at Hawkins's home, Sharpham Park, on several occasions during the early 1830s, and the comparative anatomist Richard Owen who, after visiting in 1839, described Hawkins as a 'worthy and eccentric man of genius'.[34]

When, in 1833, Hawkins announced his intention to publish a descriptive account of his ichthyosaurs and plesiosaurs, the work was awaited eagerly by the geological community. There were, however, reservations. Mantell, for instance, ordered three copies, including one for his friend Benjamin Silliman. In a letter to Silliman he predicted that the drawings would be 'beautiful', but he worried that the book would be let down by the text: 'unless Buckland (to whom it is dedicated) looks over it', he wrote, 'the letterpress will completely mar the work'.[35] Buckland received his presentation copy of *Memoirs* in June 1833, a year ahead of publication. In a letter accompanying the copy, which was reproduced in an appendix to a second edition of 1835, Hawkins wrote that he would amend the work according to Buckland's suggestions: 'I must beg you to believe that the experience I have acquired shall be applied to the correction of such indiscretions of style as now claim your indulgence, for the future, and that my most fervent aspiration is to produce such a work as shall be really worthy of your high name and approbation.'[36] Despite both of their efforts, however, there were still some complaints, the reasons for which will be explored shortly.

Memoirs was published in the summer of 1834 by the scientific publishers Relfe & Fletcher of London as an enormous folio volume, approximately fifteen by twenty inches, priced two pounds and ten shillings. It was printed on high-quality paper with huge margins, and was illustrated with twenty-eight lithographic plates. Though it was not noticed by the heavyweight quarterlies, it was reviewed in a number of literary magazines, including the *Athenaeum*, the *Literary Gazette*, the *Gentleman's Magazine*, the *New Monthly Magazine* and the *Metropolitan Magazine*. It was also reviewed in John Loudon's *Magazine of Natural History*.

With the exception of the *Gentleman's Magazine*, which had been founded early in the eighteenth century, and the *Magazine of Natural History*, the periodicals which reviewed *Memoirs* belonged to a new kind of literary magazine intended for a more general readership than that reached by the heavyweight quarterlies.[37] The *Literary Gazette*, for instance, founded in 1817 and appearing weekly at a cost of eight pence, included reviews and lengthy book sections, and was influential amongst the novel-reading public into the 1830s.[38] The *Athenaeum* appeared as a rival weekly from 1828.[39] The *New Monthly Magazine* and the monthly *Metropolitan Magazine* were founded in 1821 and 1831 respectively; the latter was notable as the first periodical to print serialized fiction.[40] In the early decades of the nineteenth century, reviews were important in communicating science to a wide and varied readership.[41] The main text of *Memoirs* would have been read only by the small number of people who were able to purchase or perhaps borrow it; these readers would mostly have been wealthy, as the book was very expensive, and many would have had some specialized knowledge of oryctology. Many more, however, would have learned about Hawkins's books from reviews. The *Literary Gazette* had a circulation of approximately 4,000 in 1823, although this may have decreased by the time *Memoirs* was published, the *Athenaeum*'s circulation peaked at about 18,000 in the late 1830s, and the circulation of the *New Monthly Magazine* was about 4,000.[42] Bearing in mind that many copies would have been read by more than one person, reviews of *Memoirs* may have reached tens of thousands of people.

Hawkins was well aware of the impact that the periodical press would have on contemporary perceptions of himself and his writings. Upon reading the *Literary Gazette*'s review, which contained some favourable comments, he immediately forwarded a copy to Buckland. Less pleasing was the review in the *Athenaeum*, quoted in the introduction to this chapter, which ridiculed *Memoirs* through the use of heavy irony. Hawkins responded with great indignation in his preface to the second edition of *Memoirs*, describing the *Athenaeum* as 'a weekly print, of which very likely the reader never before heard'.[43] This was almost certainly an instance of wishful thinking, as the *Athenaeum* was probably the most influential review of the period.

Book reviews are shaped by a variety of concerns, including, in addition to the opinions of their authors, the expectations of publishers and readers. As the nineteenth-century literary elite debated the nature of literature and of criticism, professional critics scraped a living by adhering to the fundamental principle of reviewing which Edward Copleston set out in his satirical *Advice to a Young Reviewer, with a Specimen of the Art* in 1807: '*Write what will sell*.'[44] Used cautiously, however, reviews are helpful indicators of contemporary perceptions of literary works. In the sections which follow I read the above-mentioned reviews of *Memoirs*, plus reviews of *Sea-Dragons* from the *Metropolitan Magazine* and the *New Monthly Magazine*, for clues as to how contemporary readers perceived Hawkins's publications. Working outwards from these clues, I apply techniques of close reading and, in particular, genre analysis to *Memoirs* and *Sea-Dragons* in order to explain these readers' reactions.

Anatomy and Genre

Traditional studies of genre have dealt primarily with texts, yet texts cannot exist apart from their materialized forms. As the *New Monthly Magazine*, quoted above, recognized in 1840, what meanings any book, say, can have depends both on how that book is 'written' and how it is 'put together'. The literary theorist Gérard Genette has coined the term 'paratext' to denote all that which 'enables a text to become a book and to be offered as such to its readers and, more generally, to the public'.[45] Empirically composed of a heterogeneous set of practices and discourses – some old (the title, the preface, the illustrations), others new (the printed cover) – the paratext functions as a threshold of interpretation. As the first site of interaction between reader and text, it is responsible for generating preconceptions which will continue to inform how the text is read and understood, even if those preconceptions are never met in full.[46] In his preface, Hawkins states the purpose of *Memoirs* thus, citing paratextual features of the book as evidence: 'the volume now before the reader has but modest claim – indeed, the title anticipates it ... – the assemblage of facts relative to Ichthyosauri and Plesiosauri, merely. To this end I had but to study their remains as an anatomist, and, if I may boast, that branch of science has not been neglected by me, and to watch vigilantly the progress of my plates, which are, after all that is said, the best interpreters of the original matter, if carefully examined'.[47]

Let us begin with the title. The title page of Hawkins's first fossil book is shown in Figure 3.3. Its most prominent feature is the title, which takes up just under half a page: *Memoirs of Ichthyosauri and Plesiosauri, Extinct Monsters of the Ancient Earth, with Twenty-Eight Plates, Copied from Specimens in the Author's Collection of Fossil Organic Remains*. The minimal functional requirement of a title is to designate a particular work by naming it; as Genette points out, to these ends any title will do.[48] This particular title, however, also serves additional functions. It states, for example, that the work contains twenty-eight plates; lithographic plates were expensive, and so the title serves to tell potential purchasers what they can expect for their money in material terms. It also provides information about the subject of the work. The book is about ichthyosaurs and plesiosaurs, and the plates are copied from fossil organic remains.

The title also proclaims the work to be a collection of 'memoirs'. The term 'memoirs' was commonly used in book titles during this period in a restricted sense to mean a history or a biography based on the author's personal knowledge of the subject. In the case of a historical memoir, this knowledge was often acquired first hand through participation in the events narrated; in the

MEMOIRS

OF

ICHTHYOSAURI

AND

PLESIOSAURI,

EXTINCT MONSTERS

OF THE ANCIENT EARTH,

WITH TWENTY-EIGHT PLATES,

COPIED FROM SPECIMENS IN THE AUTHOR'S COLLECTION OF FOSSIL ORGANIC REMAINS.

BY

THOMAS HAWKINS, Esq. F. G. S.

&c. &c. &c.

I believe in, and bear humble testimony to the Scriptures, that they are the "words of everlasting life" and   he wisdom of God." I cannot pretend to understand much of them, for reasons which they themselves offer, much less can I presume upon their explanation; but I owe to myself and the reader to say—to prevent any misconception of the spirit and letter of this book—that I think the Mosaic cosmogony intelligible upon this hypothesis :—*our* creation is not *a principio*, but from the fourth day or generation of Time, when the lights in the firmament were made "to give light upon the earth." The antecedent history of the planet, as written by Moses, demonstrable by the soundest physics, unshrouds but the gaunt skeleton of the pre-Adamite epoch, to the clearer comprehension of which nothing can so well serve as the accumulation of Fossil Organic Remains.

FROM THE AUTHOR'S INEDITED MSS.

LONDON:

PUBLISHED BY RELFE AND FLETCHER, 17, CORNHILL.

MDCCCXXXIV.

Figure 3.3. Title page of *Memoirs of Ichthyosauri and Plesiosauri*. Reproduced by kind permission of the Syndics of Cambridge University Library.

commoner biographical case, where the work was not autobiographical, the authority for the memoir was generally derived from the author's having known the subject personally, or from the author's being privy to oral or written communications made by the subject or the subject's close associates. The term 'memoirs' implied a certain epistemological relationship between author and subject, whereby privileged knowledge was derived from direct communication. There were precedents for the writing of memoirs within the field of geology – see, for example, Thomas Ashe's *Memoirs of Mammoth* (1806), Thomas Weaver's 'Memoir on the Geological Relations of the East of Ireland' (1821) and Peter Martin's *Geological Memoir on Part of Western Sussex* (1828) – and in his preface Hawkins explained that he chose the term to mean 'a familiar exposition of one's own ideas in a latitudinarian degree'.[49] Nevertheless, the potential ambiguities of his title were noticed by reviewers. 'The title of this work is not exactly correct', wrote a reviewer for the *Metropolitan Magazine*, wryly:

> Memoirs would imply a history or biography collected from actual knowledge, or some oral or written communication. Now where Mr. Hawkins may have obtained his memoranda, to compile the biography of pre-existent animals, is rather puzzling. It rather ludicrously occurred to us, that any old lady who was not well versed in natural history, upon reading the title, might exclaim ...'Well I wonder who can those two young gentlemen be with such hard names.'[50]

In fact, Hawkins considered an element of direct (divinely facilitated) communication with the past to be central to his work – this theme is revisited later on.

Other paratextual features of *Memoirs* had more straightforward generic implications. Inscribed in the centre of the title page, for instance, is the name of the author: 'Thomas Hawkins, Esq. F.G.S.' Hawkins was elected Fellow of the Geological Society in 1832. His application was supported by Charles Lyell, Henry De la Beche, Edward Turner, Daniel Sharpe and William Clift, a respected anatomist who worked at the Hunterian Museum.[51] To an outsider, the letters 'F.G.S.' would have conferred scientific authority upon the book. To members of the elite circle of geologists who associated with Hawkins, by contrast, the name itself would have been more significant. Hawkins received considerable support during his early career: William Buckland, for instance, lobbied the British Museum on his behalf regarding the purchase of his collection in 1834, and encouraged Conybeare, De la Beche, Gideon Mantell and Clift to write letters to the Museum backing the purchase.[52] Attitudes towards him, however, were paternal and at times condescending. Mantell, for instance, described him in private as 'a very young man, not versed in science and knowing but little of anatomy' and, on another occasion, as 'a very young man who had more money than wit, and happened to take a fancy to buy fossils'.[53] It is possible that, on occasion, gentlemanly geologists may simply have humoured Hawkins in order to gain access to his extremely significant collection. Be that as it may, he succeeded in making the most of his connections.

Memoirs had two dedications. The first part of the book, the 'Book of the Ichthyosaurus', was dedicated to William Buckland 'in acknowledgement of the author's sincere regard and most faithful devotion'. The second part, the 'Book of the Plesiosaurus', was dedicated to William Conybeare 'with the author's profound respect, and most grateful attachment'.[54] Through these dedications Hawkins was able to claim a close relationship to these two individuals and, by a kind of metonymy, the geological and scientific communities at large. Hawkins recognized the importance of the weight of his dedications. In July 1834 he wrote to Buckland, acknowledging that *Memoirs* had achieved 'a consideration that it would not have possessed but for your name on the dedicatory page'.[55]

A similar function was served by the impressive list of subscribers, shown in Figure 3.4. Of the forty-eight listed, nineteen were Fellows of the Geological Society, of whom fourteen were also Fellows of the Royal Society; these included William Buckland, William Conybeare, Gideon Mantell, Adam Sedgwick, Henry De la Beche and Roderick Murchison, six of the most eminent geologists of the day. While it was common in this period to preface a book with a list of subscribers, Hawkins's list has some unusual features. Atypically, he appears to have ordered his subscribers into an approximate hierarchy, with the Fellows of the Royal Society at the top. After these come notable female geologists and collectors, including the Wiltshire geologist Etheldred Benett (1775–1845) and a Miss Philpot, who must have been one of the Philpot sisters of Lyme Regis.[56] The women geologists are followed by those subscribers boasting fellowship of the Geological but not the Royal Society. Amongst the remainder are geologists without an institutional affiliation, such as the naturalist Walter Calverley Trevelyan, and friends and acquaintances of Hawkins, such as Thomas Porch Porch [*sic*] of Glastonbury.[57] Rather unfortunately, the list contains a high proportion of typographical errors, which cannot have helped endear Hawkins to the elite geologists whose approval he manifestly craved: 'The Rev. A. Sedgewick' rather than 'Sedgwick'; 'Phillip Grey Egerton', whose first name should have been spelled 'Philip'; 'Robert Murchison' for Roderick Murchison; and 'William Cliff' for William Clift, to pick up on just a few.[58] Nevertheless, placed prominently on the second page of the book, the list served as a genre indicator and as an advertisement. This was a geological work, of interest to scientific readers, and endorsed by prominent members of the scientific community.

Within the field of geology there were many different kinds of work on offer.[59] From its physical form alone, readers would have been able to draw conclusions about the intended audience for *Memoirs*. With its large-format, high-quality paper, wide margins and costly lithographic plates, *Memoirs* clearly did *not* belong to any of the broadly scientific genres that catered to the less afflu-

SUBSCRIBERS.

The Rev. W. Buckland, D.D. F.R.S. G.S. & L.S. &c. &c. &c.
The Rev. W. D. Conybeare, F.R.S. G.S. I.R.S.P.C. &c. &c. &c.
The Marquis of Northampton, F.R.S. G.S. &c. &c. &c.
Gideon Mantell, Esquire, F.R.S. G.S. & L.S. &c. &c. &c.   Three copies.
The Rev. A. Sedgewick, F.R.S. G.S. &c. &c. &c.
Doctor Fitton, F.R.S. G S. &c. &c. &c.
Sir Phillip Grey Egerton, Bart. F.R.S. G.S. &c. &c. &c.
Henry T. de la Beche, Esquire, F.R.S. G.S. &c. &c. &c.
Robert Impey Murchison, Esquire, F.R.S. G.S. &c. &c. &c.
J. S. Bowerbank, Esquire, F.G.S. L.S. &c. &c. &c.
William Cliff, Esquire, F.R.S. G.S. &c. &c. &c.
William Tooke, Esquire, M.P. F.R.S. &c. &c. &c.
Thomas Bell, Esquire, F.R.S. G.S. &c. &c. &c.
Thomas N. N. Morson, Esquire, &c. &c. &c.
R. H. Solly, Esquire, F.R.S. G.S. &c. &c. &c.
Miss Benett, Norton House, Warminster.
Miss Philpots, Lyme-Regis.
Miss Bailward, Frankley House, Wilts.
Miss Lousada, Bedford Place.
Doctor Moore, A.M.
Colonel Silvestop, F.G.S. &c. &c. &c.
Henry T. M. Witham, Esquire, F.G.S. W.S. &c. &c. &c.   Two copies.
G. W. Braikenridge, Esquire, F.G.S. & S.A. &c. &c. &c.
Charles Barclay, Esquire, F.G.S. S.A. M.R.A.S. &c. &c. &c.
J. Chaning Pearce, Esquire, F.G.S. &c. &c. &c.
Doctor Roberts, M.D. F.R.C.A. &c. &c. &c.
John Barker, Esquire, M.R.C.S.L. &c. &c. &c.
J. H. Curtis, M.R.I. &c. &c. &c.
Frederic Edwards, Esquire, &c. &c. &c.
James Edwards, Esquire.
Mr. Roberts, Lyme-Regis.
John Hilton, Esquire, &c. &c. &c.
Major Roach, Glastonbury.
Thomas Porch Porch, Esquire. A.B. &c. &c. &c. Glastonbury.
Orlando Reeves, Esquire, Taunton.
Mr. W. D. Saul, F.G.S. &c. &c. &c.
The Rev. B. Jeanes, Charmouth.
Augustus Langdon, Esquire, &c. &c. &c.
Colonel Macerone.
J. H. Waugh, Esquire.
Doctor P. Murray, Scarborough.
Frederic Forster, Esquire, &c. &c. &c.
J. E. Jackson, Esquire, Brasenose, College.
The Scarborough Institution.
The Rev. J. Image, Dulwich College.
William Mills, Esquire. F.L.S.
W. C. Trevelyan, Esquire, Wallington.
James Carter, Esquire.

b

Figure 3.4. List of subscribers to *Memoirs of Ichthyosauri and Plesiosauri*. Hawkins was keen to stress the institutional credentials of his supporters. Reproduced by kind permission of the Syndics of Cambridge University Library.

ent reading public. It was clearly *not* a textbook or manual like De la Beche's *Geological Manual* (1831) or John Phillips's *Guide to Geology* (1834), for example.[60] Neither was it a cheap periodical publication, such as the *Penny Magazine*, which had included articles on the ichthyosaur and the plesiosaur in its series on the 'Mineral Kingdom' the previous year.[61] Equally, it was not a religious tract on a natural-theological theme, a pamphlet, a broadside, a children's book, or a volume from an 'improving' series such as Charles Knight's *Penny Cyclopaedia* (1833–44). *Memoirs* was a luxurious, expensive book, and was thus intended to be purchased and read by members of the upper ranks of society.

As regards the branch of geology that *Memoirs* was concerned with, the most important indicator was the plates, which showed it to be a work concerned specifically with the technical representation of the anatomy of fossil saurians. The plates were large, high-quality lithographs. The medium itself would have been significant. Lithography was a favoured means of reproducing technical or scientific diagrams because it allowed fine detail to be represented. Furthermore, it was direct: images were drawn straight onto the stone without any need for engraving. However it was only suited to certain kinds of book because it was expensive. That the plates were finely crafted lithographs, rather than cheap woodcuts, for example, implies that technical detail was important to their meaning.

A diagram of *Plesiosaurus triatarsostinus* is shown in Figure 3.5. The illustration was drawn by Hawkins and lithographed by a professional lithographer named Rossitor. The plate comprises four views of the head and jaws, one view showing the sternum, shoulder and forelimbs, and a final side view of the pelvis and hindlimb of the animal. The representations are essentially linear, with only minimal shading and variation of line thickness to indicate depth. They are abstract, representing an idealized animal rather than a specific fossil: there are no omissions, for instance, even though real specimens invariably had bones missing. For each species – Hawkins identified four species each of ichthyosaur and plesiosaur based and named according to the configuration of the bones in the paddle – the same views are given in the same order. Each bone is labelled, the labels corresponding to descriptions in the body of the text. These diagrammatic conventions were standard for anatomical illustrations. Figure 3.6, for instance, is Conybeare's restoration of the head of a plesiosaur, based on a crushed skull discovered at Street. Conybeare's mode of visual representation of the anatomy of the head is almost identical to that of Hawkins.

The representational conventions employed in the anatomical diagrams indicated to readers how the plates should be viewed, and the standards by which they should be judged. 'Accuracy' was an important criterion. Mantell, for instance, wrote to Silliman, 'the plates are accurate, and the specimens quite as perfect as they are represented'.[62] Mantell was in a strong position to make

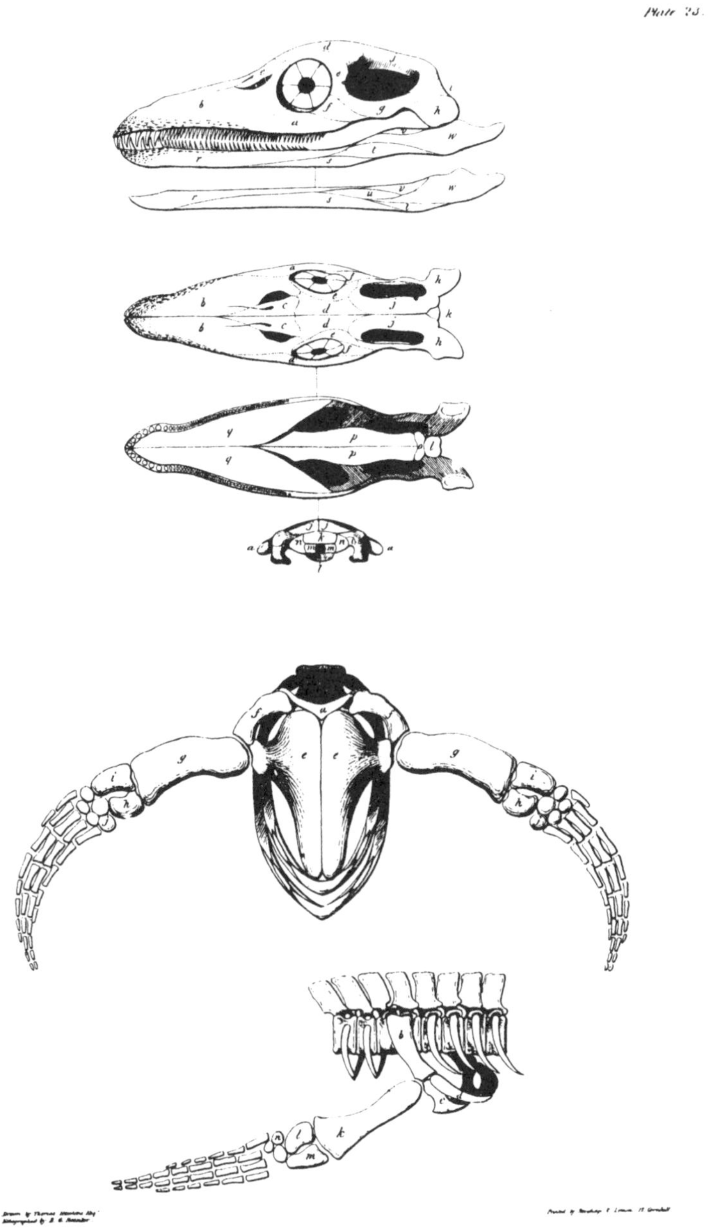

Figure 3.5. Schematic diagram of the skeleton of a plesiosaur. *Memoirs of Ichthyosauri and Plesio-sauri*, plate 23. Reproduced by kind permission of the Syndics of Cambridge University Library.

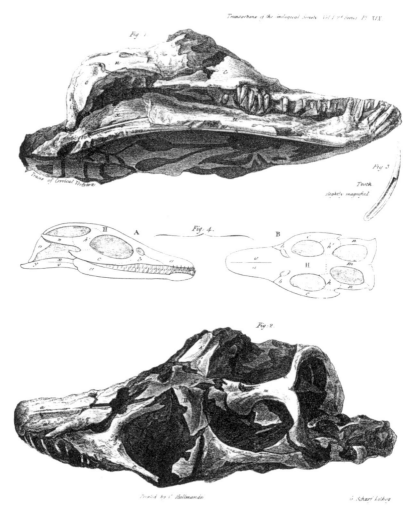

Figure 3.6. Hawkins used conventions similar to those of established palaeontologists in his schematic representations. This diagram is from William Conybeare, 'On the Discovery of an Almost Perfect Skeleton of Plesiosaurus', *Transactions of the Geological Society*, 2:1 (1824), plate 19. Reproduced by kind permission of the Syndics of Cambridge University Library.

this claim as he had viewed the specimens himself on a number of occasions. For readers not so privileged, however, accuracy was implied by stylistic features of the diagrams. A geologically trained reviewer for the *Magazine of Natural History*, for instance, observed: 'The lithographic plates ... are of large size, and well executed, and display with much clearness the osteology of the several species of fossil Ichthyosauri and Plesiosauri in the author's possession; indeed, they convey almost as distinct information as we could obtain from the specimens themselves.'[63] 'Clearness' – presumably derived from the plates' large size, their uncluttered arrangement, the linearity of the drawings, their precise labelling and so on – implicitly lent credibility to the information conveyed.

Parts of the text of *Memoirs* were written in a style characteristic of conventional anatomical description – I shall call this the 'anatomical style'. Each of the schematic illustrations is accompanied by a verbal description in the main text. These all follow a similar pattern: after a few general observations, the bones of the head, trunk and extremities (that is, the limbs) are described in turn. This was the conventional order in which to describe the bones of a quadruped: it was followed, for instance, by Cuvier in his *Ossemens fossiles*, the standard reference work on fossil anatomy in this period, and by Conybeare in his paper on *Ichthyosaurus* and *Plesiosaurus*.[64] In the case of the head, the bones are first listed, and then described individually; again the order of presentation is typical, with the exception that the teeth, rather than being described first, are described after the other bones of the head. The description of *Plesiosaurus triatarsostinus* (see Figure 3.5), for example, begins with the following list:

THE HEAD.
(SUPERIOR JAW.)
Two Maxillary . . . .*a*. . . . . plate XXIII.
Two Intermaxilliary . . . .*b*. . . . . ”
Two Nasal . . . .*c*. . . . . ”
(CRANIUM.)
Two Median Frontals. . . . .*d*. . . . . ”
Two Posterior Frontals . . . .*e*. . . . . ”
Two Jugal . . . .*f*. . . . . ”
Two Temporals . . . .*g*. . . . . ”
Two Mastoidean . . . .*h*. . . . . ”
Two Tympanals. . . . .*i*. . . . . ”
One Superior Occipital. . . . . *k* . . . . ”
One Inferior Occipital. . . . .*l*. . . . .” [etc.]

After the list, each bone is described individually in terms of its position, its shape, and its connection to other bones. For example:

THE SUPERIOR JAW.
THE MAXILLARIES. *a*. The form of the maxillaries, which are situated at the anterior and inferior part of the orbit, is that of a triangle with *an outer convex* and an *orbital surface* and *an inner region* hollowed. They have also *a base, two lateral edges* and *an apex*. The *two lateral edges* constitute, above, the anterior and inferior cilary ridge, which is very sharp and prominent; below, the external boundary of the dental fossæ which occupy nearly a half of their inferior and anterior part.

> *Connexion.* Anteriorly at their base they connect themselves by serrated suture
> with the intermaxillaries, at their posterior extremity to the jugals in the like manner
> and within and below to the palatals.[65]

The layout of the text, with its cross-references and its use of different typefaces
– capitals, small capitals, italics – in conjunction with section breaks to divide
the text into short sections, implies that *Memoirs* is a work concerned with
presenting information in a systematic, non-discursive way. Such a layout was
typical of anatomical works: Cuvier's *Ossemens fossiles* and Conybeare's 'Notice
of the Discovery of a New Fossil Animal', for example, both use font styles and
section breaks to divide up their texts, and both cross-reference relevant illustrations. The diction and syntax of the descriptions are also typical. Many of the
words belong to an esoteric anatomical vocabulary – 'Maxillary', 'Jugals', 'Mastoidean' – which would have been familiar only to readers with some knowledge
of anatomy. Others are taken from the vocabulary of geometry: 'The form of the
maxillaries ... is that of a triangle with an outer convex and an orbital surface ...
They have also a base, two lateral edges and an apex.' It was usual to describe the
shapes of bones in such terms. Conybeare, for instance, described the vertebrae
of the plesiosaur, in contrast to the vertebrae of the crocodile, thus: 'In place of
being concave at one extremity, and convex at the other end ... they are slightly
concave at both extremities of their body, but again slightly swelling in a contrasted curve near the middle of the circular area.'[66] The effect of this vocabulary
is to lend the descriptions an air of objectivity for, in theory at least, any shape
could be described in the 'neutral' language of geometry and anybody who knew
that language would understand.

The way in which the words of the descriptions are put together into sentences also distinguishes the anatomical style from other styles which Hawkins
uses elsewhere. The prose descriptions are composed of short, grammatically
complete sentences of simple construction. Where there is more than one clause
these are often connected with a simple 'and', and there are relatively few sentences in which the close of the syntactic structure is suspended until the end
of the sentence. The result of this is that the descriptions have an air of 'simplicity': they do not employ formally complex language, and thus they appear
to be devoid of rhetoric. Furthermore, the absence of any ostentatiously figurative language implies the absence of 'expressiveness'. This is approximately the
kind of writing which Roland Barthes has characterized as 'writing at the zero
degree'.[67] The author is barely inscribed in the text. The only indication of an
author in the description of *Plesiosaurus triatarsostinus* comes at the very beginning in the form of an introduction: 'We now proceed to explain the anatomy of
the Plesiosaurus Triatarsostinus'. With this 'we', the author announces the com-
mencement of the description but then disappears for eight-and-a-half pages. In
fact, the degree of authorial absence in Hawkins's descriptions is extreme even

by anatomical standards. Cuvier, for instance, sometimes introduced references to himself in order to explain how his findings related to particular specimens he had consulted, or to make it clear that his conjectures and conclusions were, precisely, *his* conjectures and conclusions. For example, he wrote of two bones in the head of an ichthyosaur specimen, 'Il me paroît clair que les deux os que je viens de décrire sont le temporal et le tympanique' ('It appears clear to me that the two bones which I have just described are the temporal and the tympanic ones').[68] Conybeare did the same. For instance, he introduced himself implicitly as a cautious observer in statements such as: 'The number of cervical and dorsal vertebrae in this animal appears to be 46.'[69] In Hawkins's descriptions, by contrast, the bones themselves are the actors: the bones 'have', they 'constitute', they 'connect themselves'. The narratorial voice seems to speak from nowhere and yet it possesses a weighty authority which derives, first, from the authority of other, similar anatomical descriptions and, second, from the authority of the adept author himself who is, despite his efforts at concealment, always implicit in the text, for it must have been written by someone.

On the basis of these descriptions readers identified *Memoirs* as an anatomical monograph. A reviewer for the *Literary Gazette*, for example, wrote: 'The anatomical descriptions ... render the work a monograph of great scientific value.'[70] Because they conformed in large part to readers' expectations of what an anatomical monograph should be like, the descriptions could be judged against the standards of the genre. The scientifically informed reviewer for the *Magazine of Natural History*, for example, judged them 'good', but wished them to have been 'more ample'. He also criticized Hawkins's nomenclature (which was not subsequently taken up).[71] The language of the descriptions had implications even for non-specialist readers. William Weddall, for instance, reviewing for the *Gentleman's Magazine*, observed, 'the anatomical analysis of the wonderful oviparous reptiles is scientific, and, we presume, accurate': although he did not have the necessary expertise to ascertain whether the descriptions were correct or not, the language itself was sufficient to convince him that they probably were.[72]

Through technical writing, Hawkins strove to establish his competency as an anatomist. However, knowledge of anatomy was not all he wished to be known for. In communicating his wider ambitions, Hawkins used alternative styles of writing, and the resulting juxtapositions threatened to undermine the assumed generic identity of his works. This tendency towards generic promiscuity, explored in the next section, was heavily implicated in critics' assessments of *Memoirs* and *Sea-Dragons* as 'singular', 'curious' and 'eccentric' books.

Mixing Genres

Nineteenth-century geology was built on the contributions of a diverse array of men and women from all ranks of society. While labourers such as George Moon and commercial dealers such as Mary Anning were responsible for much of the essential groundwork, the focus in terms of prestige was on the elite group of geologists who formed the Geological Society of London. The majority of these were gentlemen, relieved from the necessity of routine paid work by substantial independent incomes. Hawkins, however, came from the middling rather than the genteel ranks; his inheritance was not huge, and his passion for collecting ultimately led him into financial difficulties. Yet Hawkins appears to have placed great value on social status, and he recognized that geology offered a route to assimilating with the social elite. Here I suggest that, in order fully to establish himself within the elite circle of gentlemanly geologists, Hawkins felt a need to prove that he possessed not only the necessary skills, but also the appropriate character. This could not be demonstrated through anatomical description alone.

Perhaps recognizing that he would never exceed his heroes in anatomical expertise, Hawkins placed special emphasis on his unique talent for collecting. He even tried to prove it scientifically: the phrenologist Spurzheim, he claimed, once told him after examining his head that he 'possessed the largest acquisitive organ he had ever felt'![73] In the preface to *Memoirs* and elsewhere he proclaimed the superiority of his collection explicitly. 'The collection of which it [*Memoirs*] treats ... weighs more than twenty tons, occupies a superficies of two hundred feet by twenty, and, in pretension of every sort, transcends all the collections in the world.'[74] In the body of the text he developed this aspect of his character by interspersing his anatomical descriptions with romantic, pastoralized accounts of his acquisitions. Though the discoveries themselves were generally made by quarriers or commercial dealers – Hawkins acknowledges and sometimes names these individuals – Hawkins himself always plays the geological hero, differentiated from his lowlier assistants by education, class status and wealth. This kind of heroic characterization was common amongst gentlemen geologists, who derived great amusement from it. Buckland, for instance, was dubbed the 'Ammon Knight' after he galloped off from a collecting expedition with a huge ammonite over his shoulder.[75] However, while the development of such characters was generally restricted to personal correspondence and memoirs, in Hawkins's case it formed a major part of his anatomical works.

'Twas July —'. The year was 1830 and Hawkins, just twenty years old, was on the brink of an experience that would change the course of his life for ever:

> ... the day had been sultry hot – cool the delicious eve; and in a retired cottage that may be well called Virgilian – so agreeable its situation on the woody hills that overlook

the wide moors of Glaston – discoursing many pleasant things with my schoolmaster of old I sat. And our ranging thoughts alight upon geology – the new science – and I tell of the more than Arabian wonders of the neighbouring villages. And elevated with my story we agree to ramble the landscape that leads towards them.[76]

This is the beginning of the story of Hawkins's acquisition of his first ichthyosaur specimen, the 'radicle' of his collection. Rambling through the picturesque countryside with his companion, from Glastonbury to Street, from Street to Walton, the pair eventually arrive at George Moon's 'humble abode'. They are greeted by his wife, who shows Hawkins an ichthyosaur fragment found the day before. Hastening towards the local quarry they encounter Moon on his way home from work: 'Show me whence you got those pieces of marl, Moon.' Protruding from the wall of the quarry are an enticing section of rib and several vertebrae. Here is the heroic challenge: to rescue the specimen. Moon protests that it is too late to start work, but Hawkins is insistent and impatient. A team of quarriers, already worn by a long day's labour, is hired to work on the specimen through the night by candlelight. Moon is left to supervise, while Hawkins returns giddily homewards.

In his acquisition accounts Hawkins adopted an altogether different linguistic style, based not on anatomical writing but on pastoral poetry. Pastoral poems enjoyed a high degree of popularity through the eighteenth and into the nineteenth century. They drew on highly conventional, idealized images of the day-to-day activities of shepherds and other country folk in order to express an idyllic vision of the peace and simplicity of rural life.[77] Hawkins drew on stock subjects of the genre:

> And the sweet-smelling new-made hay and the happy peasant; the sunny maiden returning from the kine with pail brimful of milk. And floating the ambient air – gossamer-like came the bleating of sheep, the drone of the blind-beetle and the mellow tones of a distant fife.[78]

That this narrative was deliberately composed as a pastoral is indicated by the reference in the opening lines to Virgil, who, in his imitations of the Theocritean idylls, established the model for the genre. In a further reference to the historic precedents of the genre Hawkins depicts rural Somerset as Arcadia, the region of Greece in which Virgil set his pastorals: 'And our ravished eyes behold an expanse fairer than the fair plains of Arcady – brown moors and corn-fields unnumbered, and the Mendips far-stretched to the North and East.'[79]

This passage, like other discovery accounts in *Memoirs*, differs radically from the anatomical descriptions in terms of its style. The sentences are longer and more complex, and the passage is narrated in the active voice rather than the passive. While the anatomical descriptions attempt to deny all traces of style, this passage exhibits it ostentatiously. In stark contrast to the technical terminology, here Hawkins employs a poetic vocabulary: 'eve' is substituted for evening, 'maiden' for girl, 'kine' for cows, 'clime' for climate, 'yea' for yes, 'morn' for morning, and so on. The very first word, 'Twas', brings with it a multitude of connotations, for the word 'twas' belongs to the stock vocabulary of the pastoral genre. Compare, for instance, the opening lines of John Clare's 'Maggy's Repentance' (1821):

> Twas sunday eve the sun was out of sight
> & left the west sky with a yellow light
> While that small wind that ushers night apace
> Was fluttering cool on summers burning face[80]

'Twas July—': in point of fact the words inform the reader that the events about to be recounted took place in the month of July; but the connotations are of a hazy summer, some time in an idyllic rural past. The tone is set for an evocative, nostalgic rendering of the awakening of an innocent, unworldly young mind to the wonders of fossil geology – the joys of a wholesome, spirited, outdoor pursuit – the joys of the country in the summer.

Whereas in the anatomical descriptions the author is effaced from the text, here he is heavily inscribed within it. Sometimes he speaks in a commentative voice: 'Give me the country in summer, country folk both summer and winter and all the rest of the year.'[81] At other times he speaks as a character participating in the narrated events. At one point this particular story breaks into dialogue:

> 'We must dig it out to night, hearties.'
> 'We can't zir, 'tis too leate.' 'We will.'[82]

Here, Hawkins exhibits his mastery of dialectal writing, a literary device commonly employed in pastoral literature. He portrays himself as a character in the work, who asserts his class status (through his use of non-dialectal spoken language, in contrast to the quarrier George Moon) through speaking directly in the work. He also conveys his sense of humour and his extraordinary passion for collecting – characteristics which he perceived to be essential to the nature of the true gentlemanly geologist.

The inclusion of pastoralized discovery accounts in *Memoirs* can be understood as part of a larger project on Hawkins's part to establish his self-identity as a collector. Many of his plates, for instance, can be interpreted not simply as illustrations of the anatomy of the various species of ichthyosaur and plesiosaur, but also as testaments to Hawkins's extraordinary acquisitive and restorative capabilities. Compare, for example, Figures 3.5 and 3.7: the species depicted – *Plesiosaurus triatrasostinus* – is the same in each, but the modes of representation are very different. While Figure 3.5 uses bold, linear marks and alphabetical labels to represent the anatomy of the species in the abstract, Figure 3.7 employs

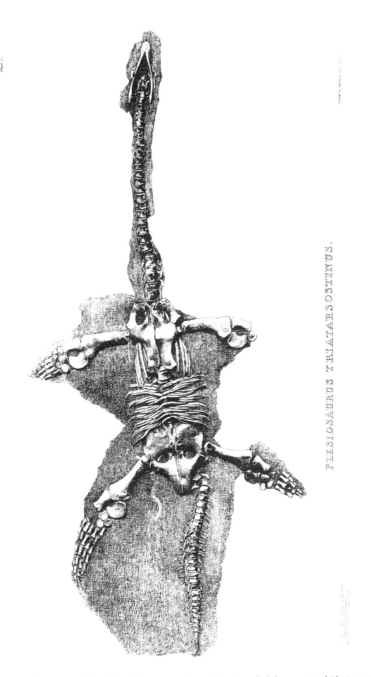

Figure 3.7. Illustration of a whole plesiosaur specimen. *Memoirs of Ichthyosauri and Plesiosauri*, plate 24. Reproduced by kind permission of the Syndics of Cambridge University Library.

Figure 3.8. Georges Cuvier's diagram is more schematic than Hawkins's portrait-like depictions of whole specimens. Cuvier, *Recherches sur les ossemens fossiles* (1821–4), vol. 5, part 2, plate 31. Reproduced by kind permission of the Syndics of Cambridge University Library.

realist conventions of shading to portray a complete, individual specimen in its tangible concreteness. While it was common enough to find whole skeletons depicted in anatomical works, Hawkins's were unusual. Cuvier, for example, included a diagram of a fairly complete plesiosaur specimen in his *Ossemens fossiles* (see Figure 3.8), but this image has very different aesthetic and moral qualities to the drawings in Hawkins's books. Cuvier's whole specimen is surrounded by numerous other drawings of fragments, presumably to fit as much information on the sheet as possible, and the drawings are all relatively linear, with depth being represented by stylized hatching. Hawkins's whole skeletons, by contrast, are skilfully executed, full-length portraits of named individuals, to be admired as much for their beauty as for the knowledge they can contribute to science. His fossil portraits elicited affective responses in non-specialist viewers: Weddall, for instance, thought they were 'the most beautifully executed delineations we have ever beheld in any geological work'.[83] They were also remarkable in scale. One drawing, which appeared in both *Memoirs* and *Sea dragons*, folded out to more than 1 metre in length. The *Metropolitan Magazine*'s reviewer of *Sea-Dragons* commented, 'its size is gigantic as its theme; its illustrations overrun fourfold the pages which hold them, and are lapped fold within fold like the enormous wings of the monsters that they represent ... In a word, the entire work is antediluvian.'[84] *Memoirs* was intended, in part, as a testament to Hawkins's technical abilities but, moreover, it was intended as a monument to his extraordinary collection. While the pastoral narratives recorded Hawkins's mighty collecting endeavours, the fossil portraits recorded the magnificent fruits of his labour.

Readers repeatedly commented on the inclusion of pastoral meditations and heroic conquests in the work. Hawkins's efforts to be recognized as possessing a sound, geological character were at least partly successful. Readers empathized with his representations of fossil collecting as a peaceful, pastoral activity. The *Metropolitan Magazine*'s reviewer, for instance wrote: 'We envy him his pursuit, we admire his talents and his perseverance, and we cannot conceive a more happy being in existence than the author seated on the Lias rock, indulging in the speculations of his ardent mind, and following up a study in which evidently consists the whole pleasure of his existence.'[85] The *New Monthly Magazine* commended Hawkins on his wit and appreciated that he had 'contrived to lead a sparkling stream of humour through the dark region of his occult researches'.[86] Other readers were more ambivalent. Conybeare, for instance, wrote privately to Buckland: 'What capital fun Hawkins's book is. I only wish it had been published before Walter Scot [*sic*] died. It might have furnished him a new character, a Geological bore far more absurd than all his other ones put together.'[87] Conybeare saw in Hawkins's self-characterization all the elements of a geological enthusiast, but he found the result absurdly contrived. In his opinion, Hawkins came across as a stereotyped 'Geological bore' rather than a genuine geologist (such as Buckland or himself).

It should be noted that in this period there were geological books containing reflective narratives which were very well received within the scientific community. Hugh Miller's *The Old Red Sandstone; Or, New Walks in an Old Field* (1841), which opens with an account of how Miller first developed an interest in geology whilst working in a quarry, is a case in point.[88] There was, however, a significant difference between Hawkins's and Miller's books, which may account for why the former were perceived to be eccentric whilst the latter were not. Miller wrote explicitly for a 'popular' audience, while *Memoirs* was, as I have shown, framed primarily as a specialist monograph. In *The Earth on Show: Fossils and the Poetics of Popular Science, 1802–1856*, Ralph O'Connor argues that 'generic fluidity' was considered more acceptable in popular geological works than in specialist treatises. He links this acceptance to the rise of periodical miscellanies, noting that many of Miller's books first appeared in periodicals in serialized form. 'The need to churn out short essays on diverse subjects', O'Connor writes, 'meant cultivating a range of styles, leaping nimbly from topic to topic, switching tone where necessary, sprinkling anecdotes, jokes, and quotations where appropriate, and building a strong rapport with his readers by developing a personalized narrative voice.'[89] In popular geological works, a certain amount of discontinuity could be tolerated, indeed it might even be beneficial insofar as it helped to hold the attention of readers. The same was not true, however of specialist scientific treatises. Accordingly, favourable comments tended to come from non-specialist reviewers writing for literary magazines aimed at a general audience. Even here there were doubts, however: Weddall wondered if these 'decidedly singular and humorous' portions of the work were 'attributable ... to the Shaksperean [*sic*] dreams of youth, and too great a desire to soften down the asperities of science, and render his work acceptable to the general reader'.[90] Geologically trained readers, familiar with consulting technical monographs simply in order to find out about fossil anatomy, were less charitable. The *Magazine of Natural History*, characterizing Hawkins as a self-styled 'fossil Nimrod', complained that the conversations in the Somersetshire dialect were 'more amusing than instructive, and altogether misplaced in a work on science'.[91]

In the introduction to this chapter, I showed that many readers struggled to comprehend Hawkins's books. One possible reason for this is that, in liberally mixing diverse literary styles, Hawkins filled his writings with spatial and temporal discontinuities. The conventions of the anatomical style are such as to make the descriptions appear placeless and timeless: they generally refer to abstract, ideal specimens which have no concrete physical or temporal reality. The acquisition accounts and other autobiographical fragments, by contrast, are associated with particular, familiar locations – the village of Walton, the coast at

Lyme Regis, and so on – and are temporally located in a precise, recent, familiar past. Hawkins's visions, discussed in the next section, are different again, for while they are geographically and temporally located to some degree, these locations are imprecise and difficult to conceive of. Hawkins's ichthyosaurs and plesiosaurs may have died and been buried in the region of the Earth now known as Somerset, but the landscape, climate and moral quality of that Earth, as Hawkins describes it, are so alien as to be almost unthinkable. From page to page, paragraph to paragraph, Hawkins's reader is asked, implicitly, to adjust to these changes.

The following extract from *Memoirs*, for instance, follows immediately on from the pastoral set-piece just discussed:

> The solemn bat beat the air with membranous wing as I journeyed lazily to Glastonbury so – meditating – methought I glided adown the stream of time into the oblivious profound where flit the unreal shadows of extinguished generations. And amid the horrible darkness I beheld monstrous and dire skeletons that came from nature's untutored hand ere the Divinity called order from chaos forth. And shapes big as leviathan and more terrible, and ghosts of unfinished existencies [*sic*].[92]

The narrative is continuous with the preceding passage – Hawkins is walking back to Glastonbury having left George Moon to supervise the extraction of *Ichthyosaurus chiroparamekostinus* – and yet within the first sentence the reader is snatched from the now-familiar landscape of Glastonbury and plunged into an uncanny, geographically-ambiguous darkness. The 'oblivious profound' and unreal shadows of extinguished generations do not belong to pastoral literature. We are not in Arcadia any more. Where we are is in an imaginary space of a different kind – a waking dream, a 'vision', of time when ichthyosaurs and plesiosaurs and other hideous monsters ruled the chaotic Earth. Hawkins's speculations about the very early history of the Earth were centrally important to the eccentric reputation of his books, as we shall soon see.

Genesis, Geology and the Perils of 'Enthusiasm'

Figure 3.9 shows the frontispiece to *Memoirs*. Hawkins designed the image himself and had it drawn by the landscape painter John Samuelson Templeton. In the foreground an ichthyosaur basks idly under a palm tree, while another paddles aimlessly in the ocean; a plesiosaur waddles towards the shore to greet its mate and, a little further in the distance, a pair of pterodactyls soar over the still water. While Hawkins's schematic diagrams represented the anatomy of abstract specimens and his figurative full-length plates portrayed individuals as they appeared, restored, in 1834, the frontispiece depicts the world as it might have appeared when ichthyosaurs and plesiosaurs roamed the land. It is, to use Martin Rudwick's term, a 'scene from deep time'.[93]

Figure 3.9. Frontispiece to *Memoirs of Ichthyosauri and Plesiosauri* showing Hawkins's vision of the Earth before ichthyosaurs and plesiosaurs became extinct. Reproduced by kind permission of the Syndics of Cambridge University Library.

Scenes from deep time constituted a well defined genre by the time *Memoirs* was published. Reconstructing scenes from the past had a long tradition within historical painting, and during the early decades of the nineteenth century such representations became the stock material of popular shows and spectacles.[94] Until the 1830s historical paintings depicted only scenes from human history and mythology. The first serious attempt to depict the prehistoric world was made by the geologist Henry De la Beche in 1830. Prints of his 'Duria antiquior', (An Earlier Dorset), were sold for the benefit of Mary Anning, who was suffering from financial hardship at the time.[95] The drawing was initially circulated amongst wealthy collectors, but the scene soon became famous in geological circles. Imitations followed and within a few years similar scenes were appearing in the popular press. In 1833, for example, a wood engraving depicting a plesiosaur, an ichthyosaur and other Liassic animals in a naturalistic setting was printed by the *Penny Magazine*, in an article entitled 'Organic Remains Restored'.[96] As Rudwick observes, Hawkins's animals in his frontispiece appear to be derived from De la Beche's drawings. However, while De la Beche's design is crowded and lively, Hawkins's is intended to represent the bleak emptiness of the pre-human Earth.[97] 'Theirs was the pre-Adamite – the just emerged from chaos – planet', he explains in the body of the text, 'through periods known only to God-Almighty: theirs the eltrich world uninhabitate, sunless and moonless, and seared in the angry light of supernal fire'.[98]

Hawkins's frontispiece represents, in a sense, the ultimate goal of his researches. This goal is spelled out in an inscription on the title page (see Figure 3.3). The inscription begins, 'I believe in, and bear humble testimony to the Scriptures, that they are the "words of everlasting life" and "the wisdom of God".' It affirms the author's full acceptance of the account of Creation as told in Genesis, assuming that time is taken to begin on the fourth day, with the creation of the heavenly lights. 'The antecedent history of the planet' – that is, before the creation of the Sun and Moon – 'as written by Moses, demonstrable by the soundest physics unshrouds but the gaunt skeleton of the pre-Adamite epoch, to the clearer comprehension of which nothing can so well serve as the accumulation of Fossil Organic Remains.' Hawkins believed that his fossil skeletons descended from this age before man and before the beginning of time. His goal was to use his collection to clarify those passages of Scripture dealing with the earliest history of the Earth, and flesh out the skeletal narratives embodied within them. He believed he had privileged access to this remote period in Earth history thanks to his unique collection, which in turn equipped him with a unique visionary insight into the past.[99]

Hawkins viewed his task as akin to that of the antiquarian, where his materials were not man-made artefacts but geological specimens. The analogy between geology and antiquarianism was commonplace in this period.[100] Charles Lyell, for example, in the opening pages of his *Principles of Geology* (1830–3) wrote: 'minerals, rocks, or organic remains … when well examined and explained, afford data to the geologist, as do coins, medals, and inscriptions to the historian'.[101] Hawkins reiterated this position throughout both of his fossil books, formulating it most explicitly in *Sea-Dragons*. At the beginning of a chapter on *Ichthyosaurus chiroparamekostinus*, he likened his own endeavours to those of Humphry Davy, a great man of science – President of the Royal Society, no less – but also an antiquarian:

> I now beg to present my Reader the most perfect of all the Annals belonging to Ichthyosauri. The labors by which we are enabled to recover and perpetuate them are not unlike those of Sir Humphrey [*sic*] Davy, over the papyri of the long entombed Cities of old Latium. Striving against the destructive weapons of Time, of fire, of earth, of air, and water, with the most delicate tests and a finished acumen, he was enabled to save the Hearts of some Cinerous Rolls of Herculaneum and Pompeii, so marvellously transmitted to these latter Days.[102]

During the 1820s, Davy had undertaken a series of investigations, applying chemical means to the task of unrolling carbonized papyri found at Pompeii and Herculaneum. Hawkins saw himself as engaged in a similar project: he too was engaged in interpreting 'annals' or historical records; he too had to strive and labour against the destructive weapons of time to lay bare signs from the past and allow them to speak to the present. Like Davy, Hawkins faced innumerable difficulties. The papyri and the fossils, he explained, had 'both lost their cases and externals, by the indisposition and contempt of Time on the one side, and his officiousness on the other'. Furthermore, the very processes employed in recovering the knowledge embodied in these ancient artefacts necessarily contributed to their degradation: 'In cutting the hard, intractible limestone from the involved Skeleton, how many muscles, nerves, nay, the Sensorium in which life couched and subsisted, is destroyed, as effectually as the wisdom and wit of man for ever lost in those lamented Scrolls.'[103]

Hawkins found himself constantly frustrated by the jealous resistance of time, yet it was, he argued, precisely the difficulty of his enterprise which ultimately made it so noble:

> The sublime discloses itself only in the silence of which we speak, when, by the most stupendous Efforts of Intellect, by the revivification of Worlds, by the inhabitation thereof of all the Creatures which the laboring Soul can re-articulate, we stand in a Presence which has not, nor ever shall have one sympathy with ourselves.[104]

Presented with only the tiniest of clues about the outward form of his extinct reptiles – 'the stains of the once living flesh and blood, in the softened and discolored stone' – Hawkins was able, he claimed, through 'the most stupendous Efforts of Intellect', to effect a 'revivification of Worlds'.[105] In emphasizing the work required, Hawkins placed value on his work as a fossilist, which required painstaking care and effort to reassemble complete specimens from the fragmented remains that were recovered. But he also placed value on his work in reanimating, first imaginatively and then through language, the past worlds, and the creatures which belonged to them, that were so faintly described in the Mosaic account of Creation.

In the early decades of the nineteenth century, the relationship between earth history, geology and Genesis was at the heart of debates over how geology was to be defined.[106] On the one hand, self-styled 'scientific' or 'philosophical' geologists, many of whom were associated with the Geological Society of London, increasingly asserted geology's independence as a scientific discipline: the most extreme agreed with Charles Lyell that 'The physical part of geological inquiry ought to be conducted as if the Scriptures were not in existence.'[107] On the other hand, so-called 'Mosaic' or 'scriptural' geologists – most of whom did not engage actively in geological fieldwork – reacted by asserting their 'commonsense' convictions that Genesis embodied an authoritative account of the Earth's history which could be interpreted aside from scientific geology's 'fanciful' and esoteric theories. Some examples of scriptural geology from the period include *A Short Introduction to the Study of Geology; Comprising a New Theory of the Elevation of the Mountains and the Stratification of the Earth: In Which the Mosaic Account of the Creation and the Deluge is Vindicated* (1817–19) by the Methodist clergyman Joseph Sutcliffe; Granville Penn's *Comparative Estimate of the Mineral and Mosaical Geologies* (1822); and George Bugg's *Scriptural Geology; Or, Geological Phenomena Consistent Only with the Literal Interpretation of the Sacred Scriptures, upon the Subjects of the Creation and Deluge* (1826–7).[108]

Within 'scientific' geology itself, there was ongoing disagreement over the extent to which geologists should enter into cosmological theorizing. Buckland, for instance, was criticized when, in his natural theological treatise, *Reliquiae diluvianea: Or, Observations on the Organic Remains Contained in Caves, Fissures, and Diluvial Gravel, and on Other Geological Phenomena, Attesting the Action of an Universal Deluge* (1823), he argued that his research on a supposedly antediluvial hyena den provided physical evidence that a Flood, like that described in Genesis, had occurred in recent history.[109] Martin Rudwick has argued that scientific geologists were particularly harsh in their criticism of works by members of their own social circle which 'transgressed the tacit boundaries of the science'. While in public they derided scriptural geology, he suggests that in private they considered it a serious threat to their own enterprise; as they struggled to establish their superior cultural authority, boundary maintenance was crucial.[110] More recent analysis of the relationship between scientific and scriptural geology has suggested that, in fact, there was more overlap between the two approaches than previously supposed, particularly in the case of geologi-

cal works written for non-specialist audiences.[111] *Reliquiae diluvianae* is a case
in point, for while some geologists did not like it, they did not label it 'eccentric' – it was marketed as a work on natural theology and thus its references to
sacred history were at least appropriate to the genre.[112] Nevertheless, in the case
of papers and books intended to be read primarily by other 'scientific' geologists,
scriptural topics were not considered suitable – Buckland, for example, refrained
from entering into exegetical debates in his geological papers of the 1830s and
1840s. Packaged as a specialist anatomical monograph, *Memoirs* might have
been expected to exclude biblical exegesis, just as it was expected to exclude pastoral meditations; its failure to do so contributed to contemporary assessments
of it as an eccentric book.

I suggest, however, that even more significant is the language in which
Hawkins's visions of deep time were conveyed. In his role as mediator between
past and present, Hawkins claimed to rely not only on intellectual exertion from
within, but also on 'enthusiasm' – on frenzied inspiration from without. He
explained:

> Enthusiasm seems as extinct as the Sea-Dragons which here inspire it: their strange
> eloquent Remains bespeak a Chord in our breast, which vibrates only to the Mas-
> ter Touch: the subtle and jealous gods of the vast Promontory of Time start at the
> well-known sound, They seize, They seize me wholly, and if the oracle, O Reader, be
> ambiguous, blame thy Fortune in escaping the Pythonic furor, with its extatic but
> exhausting delirium, its shiver, and wild excentric fate.[113]

'Enthusiasm' was an ambiguous term in this period, as the *Literary Gazette*'s
review of *Memoirs* illustrates: 'after such a meritorious performance, which must
have required much labour, expense, and enthusiasm, we hope that the last will
not in future be allowed to mar the general character of the author's productions'.[114] This author uses 'enthusiasm' in two senses. Insofar as it is, along with
'labour' and 'expense', one of the prerequisites for the production of a valuable
scientific work, 'enthusiasm' is intended in the positive sense of a passionate
eagerness for the pursuit of science based on a strong conviction of the worthi-
ness of the scientific enterprise. However, the original meaning of enthusiasm is
the condition of being possessed by a god, and in the early nineteenth century
this meaning was still current. As we saw in the Chapter 2 enthusiastic religion
continued to flourish in a variety of forms, especially amongst the lower classes,
but attitudes towards enthusiastic religion varied immensely. In this quotation
from the *Literary Gazette*, 'enthusiasm' is being used not only in the positive
sense described above, but also in a second, negative sense, to connote ill-reg-
ulated religious emotion and a misplaced belief in one's own privileged access
to divine communication. For many readers, Hawkins's literary 'enthusiasm'
resulted in an ambiguous text which defied all attempts at interpretation.

The ambiguity of Hawkins's visions are in part attributable to their structural complexity. In relaying his visions, Hawkins tended to use long, rambling sentences, grammatically complex or even incomplete. Furthermore, he frequently employed archaic words and constructions: 'methought', 'bespeak', 'cinerous', 'old Latium', 'ere the Divinity called order from chaos forth'. His use of capitalized nouns in *Sea-Dragons* was also self-consciously archaic: 'Chord', 'Remains', 'Master Touch'. Other, more subtle qualities added to the ambiguity of the visionary passages. For example, the obscurity of the passage cited above derives in part from slippages in the relationship of 'implied author' to text. According to a distinction originally made by Wayne Booth, the implied author differs from the actual author in that he or she exists only in the work; similarly, the implied reader is created by the work and, in Booth's scheme, functions as an 'ideal' interpreter.[115] At the beginning of the sentence quoted above the implied author assumes a commentative voice: 'Enthusiasm ... seems as extinct as the Sea-Dragons which here inspire it.' Yet while we as readers are still struggling to imagine Hawkins as the last remaining member of a near-extinct race, something strange begins to happen: by the middle of the sentence, the author seems to be narrating something which is happening to him as he writes. 'They seize, They seize me wholly ...': the delirious repetition is the utterance of someone in the very throes of a frenzied fit of inspiration. The relationship of the implied reader to the text is no more secure. Hawkins writes, 'if the oracle, O Reader, be ambiguous ...': the reader may be frustrated in his attempts to interpret the text, and yet the 'if' implies an 'if not'. The reader may be seized with a 'Pythonic furor' akin to Hawkins's, and the text's meanings may be revealed.

Many precedents for obscure visionary communications were available to Hawkins. In this passage, for example, he refers explicitly to the prophetic activities which, according to Greek mythology, occurred at Delphi. The Oracle at Delphi on mount Parnassus belonged to Apollo, god of prophecy (who, notably, had first to slay the guardian of the Oracle, the monstrous serpent Pytho, before claiming it as his own). Supplicants would ask questions of Apollo through the medium of the priestess of the Oracle, the Pythia, who was in a drug-induced, trance-like state; her responses had in turn to be interpreted by priests. In much of *Memoirs* and *Sea-Dragons*, Hawkins portrayed himself as an *interpreter* of obscure (fossil) signs, much like one of the priests of the Oracle. Here, by contrast, he is more like Pythia: a generator of oracles, which must yet be interpreted.

In Chapter 1, I discussed John Stuart Mill's essay of 1831 on the spirit of the age. Mill wrote that, 'Mankind have outgrown old institutions and old doctrines and have not yet acquired new ones', and he called for eccentric visionaries – individuals who saw things differently – to lead the nation into the future.[116] At times, Hawkins appears to cast himself in such a role. At one point in *Memoirs* he reflects on the present state of mankind in the following terms:

> But we inherit with the advantages of civilization a host of prejudices – the offspring of many incongruous ages. These, probably wholesome at their original, grow anti-quated with time and become even pernicious by aggregation. Yet other strange combinations of them are brought forth of the national genius until an unshapely idol – like the cold stalagmite formed by infinitesimal drops of water from the roof of an antediluvian cavern – is set up, before which every knee must bend and lip confess.[117]

Apparently comparing himself to the prophet Daniel (who was also familiar with dragon-like creatures; according to mythology he slew the dragon Tannin, a false prophet), he continues:

> But a Daniel belongs to every age – one whose bosom nurses so heavenly a fire that mankind acknowledges – a Titan confest. These, not squared to the dimensions of ordinary mortals, spurning the yoke which chains them to their cabined self, wander the earth exiles forbid the charities that make it habitable and descend to the tomb shriven by no friendly hand, unmourned and desolate. Oh fatal error! as though the man were the mind's inhuman Frankenstein.[118]

Hawkins did not style himself as a visionary in the Martinian mould: his middle-class background and grander social ambitions, his love of traditional learning and his academic aspirations, did not dispose him to drawing on working-class, millenarian prophetic traditions; instead he drew on biblical and classical prec-edents in constructing his visionary authorial persona. Ralph O'Connor has observed that a 'prophet-persona' lurked behind many elite geological treatises of the period, including those by Lyell and Mantell, but that Hawkins was com-paratively 'over the top' in this respect.[119] In the passage cited above, he styles himself as a wanderer, a cosmic traveller, completely free, soaring above the ordi-nary mortals of the age. In contrast to the jovial heroic geologist portrayed only paragraphs away, we have here a Promethean figure, tortured in his loneliness, his banishment from the world.

Many twentieth-century commentators, as I noted in the opening of this chapter, have read passages such as these as being utterly nonsensical. Contempo-rary readers, by contrast, recognized the visionary qualities of Hawkins's writing immediately. He 'writes like one possessed', observed one reviewer of *Sea-Drag-ons*; 'He is rapt, inspired with his subject, and his language has all the sublime of the Delphic obscurity', remarked another.[120] Some reviewers commended his literary talent. 'We wish Mr. Hawkins had written more, that he had given full and uncontrolled vent to those thoughts and feelings with which his mind is so evidently replete ... His language is poetry, and sometimes approaches to the sublime.'[121] Others, however, saw in his language traces of a diseased mind. A reviewer of *Sea-Dragons* worried that, 'for many a tedious night these same great Sea-Dragons must have sat upon his slumbers like so many hideous incubi, and have given that distortion to his language which makes the sublime so nearly approach to the ridiculous'.[122] More explicitly, the *New Monthly Magazine*'s reviewer of *Memoirs*, though approving of the book overall, noted, 'The natural consequence of an imagination so far heated, and a mind left to run wild after its favourite pursuit, is, that Mr Hawkins gives such free way to his philo-saurian bias, that we are led at times seriously to consider in what light the extent of its influence upon its possessor would be viewed by the unphilosophical High Court of Chancery in the event of his appearing before them under the rather unpleasant introduction of a Commission of Lunacy.'[123] Hawkins's obscure language presented acute critical challenges even to his contemporaries. Was it sublime or was it ridiculous? Was it poetry or was it the nonsensical rambling of a lunatic? And what, if anything, did it mean? In addressing these questions, critics reflected upon the place of rules in relation to the task of literary composition.

Hawkins further indulged his visionary propensities by publishing epic verse. His most immediate precedents were the works of Milton and his nineteenth-century followers, the so-called 'hyper-Miltonic' poets.[124] In 1840, for example, Hawkins published a poem called *The Lost Angel*.[125] The subject of the poem was the fall of angels and of man. It was received with distaste. The *Metropolitan Magazine* accused Hawkins of constructing a narrative, 'contrary to revelation, and involving much anachronism', which in places – its account of the eponymous Angel's seduction of Eve in the garden of Eden, for instance – seemed to verge on blasphemy.[126] Moreover, the reviewer complained, the poem was not well constructed. Many of Hawkins's lines were faulty in their metre, and in the very first couplet he dared to substitute a noun for a verb, which decision was judged 'not so much an effort of originality, as a specimen of bad taste'.[127]

The problem with Hawkins's poetry, according to this reviewer, was that it lacked prudence. Convinced of his own brilliance, Hawkins cast off all formal conventions and gave free rein to imagination. The results were disastrous: 'he has endeavoured to soar beyond Milton; his flight has proved lofty, irregular, and, dazzled by the height to which he had attained, has become bewildered, fallen prostrate, and failed'.[128] Like the lost Angel himself, Hawkins had broken all the rules, and the *Metropolitan Magazine* reviewer took it upon himself to punish him for his sins. The Reverend John Mitford, writing for the *Gentleman's Magazine*, was similarly displeased by the poem. Hawkins's failure, he argued, resulted 'from his having deviated from the proper track': he had chosen a 'strange wild subject' and had tried to communicate it 'in vague obscure language, and in a rough, eccentric, and abnormal versification'.[129] Mitford joked that the lost Angel was probably little used to composing poetry, and advised him to read Aristotle or William Crowe on English versification.[130]

Undeterred, Hawkins tried again. *The Wars of Jehovah, in Heaven, Earth, and Hell* occasioned less outrage, but its formal qualities were subjected to similar criticisms.[131] Mitford, again writing for the *Gentleman's Magazine*, poked fun at the author for adding 'variety to his poem by adopting new quantities in his words, as Sejánus, Mimósa, Philoctétes', affecting astonishment that he did not give 'a greater emphasis' to his own name by accenting it 'T. Hawkíns'.[132] He announced that there were 'many passages in this poem that we do not, after all our endeavours, understand' – conceding, presumably with some irony, 'It is difficult to be very sublime without being also a little obscure' – and recommended that if a popular edition be produced it be accompanied by 'a running commentary, like that in the Delphin classics, and some Scholia'.[133] In a review in the *Athenaeum*, John Abraham Heraud, himself an ambitious epic poet, admitted some poetic talent on the part of the writer, but discovered an intolerable level of 'chaos' and 'confusion' in the work. He explained: 'The poet is an artist, subject himself to rules, and imposing such upon his work. Imagination, divorced from reason, may startle by the eccentricity of its productions, but can never attract by their beauty; such births are monstrous, and, even where they compel our wonder, fail to command our praise.'[134] Like *The Lost Angel*, *The Wars of Jehovah* defied the generic conventions governing the production of epic poetry. Eccentric and monstrous it did not even begin to fulfil readers' expectations, and it thus exempted itself from the possibility of interpretation and judgement: 'in matter and manner', Heraud wrote, it lay 'beyond the pale of ordinary criticism'.[135]

In his fossil books, as in his poetry, Hawkins self-consciously rejected generic rules and constraints in favour of a personal style which expressed his innermost thoughts and feelings. In the preface to *Memoirs* he wrote: 'I speak only the language of the heart. It will offend the fastidious taste; it may even militate against some of the conventional forms which the literary world has agreed to respect, but it bears the imprint of truth.'[136] By the early decades of the nineteenth century, the old Aristotelian notion that different styles were appropriate to different genres was being challenged. During the Renaissance this idea had been formalized in the doctrine of 'decorum', according to which epics, for example, were to be written in the high or 'grand' style, while satires required the low or 'base' style. The doctrine remained popular through the eighteenth century but with the rise of 'Romantic' conceptions of authorship it began to be argued, with respect to poetry and drama in particular, that style was a unique imprint of the writer's character and so could not be subject to artificial rules or conventions. Nevertheless, many of Hawkins's critics, adhering to more traditional notions of literary decorum, were left bemused, appalled or floundering in confusion as they tried to interpret his visionary writing.

The difficulties readers experienced in interpreting Hawkins's 'eccentric' literary productions are nicely illustrated in Lord's satirical review of *Memoirs*. In his original draft of *Memoirs*, Hawkins included an unfortunate anecdote about a youthful love affair with one of his father's serving girls, identified only as 'pretty little P—'. As Hawkins explained in the preface to the second edition, the anecdote was removed during last-minute revisions, but somehow or other the *Athenaeum* got hold of it and Lord had a field day. Parodying Hawkins's assertion that geology dealt with the most sublime of intellectual challenges, with 'mysteries that require a thousand years for their solution', Lord pronounced this episode a mystery of a similar kind, requiring 'a special revelation from the author, of his own meaning'. After a whole column's worth of grappling with P—'s significance in the context of the work he found himself none the wiser: 'We have not even ascertained', he wrote, 'whether she was an Ichthyosaurus or a Plesiosaurus, whether she had "three bones in her paddle", which would have constituted her a Triatarsostinus in Mr. Hawkins's system, or possessed "the head and bill of a snipe, with two hundred and sixty long, sharp teeth", such as the Chirostrongulostinus.'[137] Presented with what appeared to be an anatomical monograph, Lord attempted to read the passage about little P— as an anatomical description. Rather than recognizing Hawkins's lover as a pretty young girl, he construed her as a fossil saurian or, worse, an uncanny human-saurian hybrid. The review was a joke, but the lesson was a serious one: generic impropriety leads always to monstrosity.

Nineteenth-century discourses of eccentricity were underpinned by ideas about taxonomic order, and the order of books was just one area in which hybridity, transgression and monstrosity were discussed and debated in such terms. Hawkins's ichthyosauri and plesiosauri were themselves described as 'monsters' and, by virtue of their seemingly aggregate nature – recall Cuvier's description of *Ichthyosaurus* – had much in common with John Timbs's modern eccentric animals and the human monsters, such as the hermaphrodite Madamoiselle Lefort, collected and displayed in the paper museums of eccentric biography discussed in Chapter 1. Like eccentric books, eccentric animal and human specimens excited wonder, disgust and curiosity in their audiences through their juxtaposition of apposite forms and natures, and aroused no faint degree of suspicion in audiences well aware of the benefits to be obtained from exhibiting the extraordinary at a price. It is to cultures of display that I now turn.

4 ECCENTRICITY ON DISPLAY: VISITING CHARLES WATERTON, TRAVELLER, NATURALIST AND CELEBRITY

The early nineteenth century witnessed an explosion of possibilities for amassing new kinds of experience: through books, magazines and newspapers, which reached out to ever wider audiences in this period; through travel, which became faster, more comfortable and more affordable; and through displays of the world's treasures and novelties at museums, exhibition halls, galleries, theatres, gardens and menageries.[1] In the midst of all these possibilities, individuals expanded their range of experiences vicariously, collecting portraits and biographies of the eminent and privileged and devouring images and descriptive accounts of far-flung corners of the world. They also acquired new experiences in person, setting out to confront the strange, the new, the exotic, the famous and the eccentric face to face.

In September 1847, William Kinsey, Rector of Rotherfield Greys in Oxfordshire, set out to do just this. In a published account of his visit to Walton Hall, home of the naturalist Charles Waterton (1782–1865), he explained:

> I had long entertained an earnest wish to become personally acquainted with the benevolent and distinguished proprietor of the far-famed Walton Hall, near Sandall, in the west riding of the county of York. I had learned to revere the character of the amiable and learned author, from frequent perusal of his admirable 'Essays on Natural History', as likewise the captivating account of his 'Wanderings in South America'. I was equally desirous to visit the house and museum of Mr. Waterton.[2]

Kinsey had the pleasure of joining 'an intellectual party' in paying a visit to Waterton at his home and, he assured readers of the *Gentleman's Magazine*, 'it was ever after a "red letter day" in my remembrance'. In fact, Kinsey was just one of many thousands of people to visit Walton Hall during Waterton's lifetime. Visitors from all walks of life recorded the intense anticipation of setting foot into the picturesque park, where birds and animals were protected against poachers and other predators; of viewing the 'museum' in the hallway, with its shimmering, lifelike specimens and satirical taxidermic monstrosities; and, most of all, of coming face to face with the 'singular and eccentric naturalist' himself.[3] In some cases the visit as it finally materialized surpassed all expectations, in others it left a faint but lingering taste of disappointment. But a trip to Walton Hall was never dull. Whether it lasted a month or an afternoon, it was bound to be full of surprises.

The Watertons were an ancient and once-illustrious Roman Catholic family who had acquired the estate at Walton, near Wakefield in the north of England, in the fifteenth century. After the Reformation, the penalties placed on Catholics caused the family's fortunes gradually to decline, a grievance which Charles Waterton never let lie. Born at the family seat, Walton Hall, in 1782, Waterton was educated at the Jesuit college of Stonyhurst in Lancashire. Upon completing his studies, he elected to continue his education through travel, since Catholics were not allowed to matriculate at the Universities at this time. In 1804 he travelled to Demerara in British Guiana, where he administered estates belonging to his father and uncle until 1812, returning home at regular intervals. As a Catholic, the traditional aristocratic professions of Oxbridge, the military and the judiciary were barred to him, so instead he chose to pursue the interest in natural history which he had first developed as a child. After 1812 he made four further voyages to the New World to collect natural history specimens, which he displayed inside his home. Over the years, he contributed more than sixty essays to John Loudon's *Magazine of Natural History*; he also published an account of his travels in 1825 as *Wanderings in South America, the North-west of the United States and the Antilles, in the Years 1812, 1816, 1820 & 1824.*[4] In 1829, by now in his late forties, Waterton married seventeen-year-old Anne Mary Edmonstone. Anne's father, Charles, was a close friend of Waterton from Guiana; her mother professed to be the daughter of an Indian princess. The marriage lasted under a year – tragically, Anne died shortly after the birth of their son, Edmund – but, at Waterton's request, Anne's sisters, Eliza and Helen Edmonstone, came to live with him, keeping his house and overseeing the upbringing of his son. From the 1830s onwards these four constituted the family living at Walton Hall.

Waterton is fairly well known today for his eccentricities – more so than Martin or Hawkins – probably because he is one of the stars of Edith Sitwell's classic, *The English Eccentrics* (1933).[5] He has been the subject of several biographies. The first, *Charles Waterton: His Home, Habits and Handiwork*, was written by his friend Richard Hobson, and published the year after his death; the latest, Brian Edginton's *Charles Waterton: A Biography*, appeared in 1996. Since Hobson, with the odd exception, biographers have tended to emphasize Waterton's eccentricity, retelling surprising adventures from *Wanderings in South America*.[6] The most famous anecdote culminated with Waterton seated uncomfortably on the back of a ferocious alligator on the banks of the Essequibo river in Guiana. His friend Edwin Jones depicted the escapade in the oil painting shown in Figure 4.1. Edginton's assessment of Waterton's 'eccentric' capers is fairly typical:

Figure 4.1. In the oil painting *Charles Waterton Capturing a Cayman*, Edwin Jones also portrays many of the other birds and animals which Waterton captured and preserved during his travels. They look on as observers, framing the picturesque scene. Reproduced by kind permission of the Governors of Stonyhurst College.

He rode an alligator, boiled a toucan, talked to insects, fought with snakes, apostrophised woodpeckers, phlebotomised himself, offered his toe to the vampire bats and indulged in all manner of scientific monkey business with the primates ... All of which and much more, advanced as evidence of eccentricity, makes a monkey out of any man, or woman, who claims that Charles Waterton was commonplace.[7]

Since the 1990s, Waterton has begun to attract the notice of academic historians. Some have analysed Waterton's hybrid taxidermic creations as manifestations of his eccentric persona.[8] A favourite has been the notorious 'Nondescript', which, with its uncanny combination of human and ape-like features (Figure 4.2), was one of the great attractions at Walton Hall, and also featured as the frontispiece to *Wanderings in South America*. Alligator riding and creative taxidermy both come into this chapter, but my primary focus is not on Waterton's activities as 'evidence' of his eccentricity, or on his personal motivations for his supposedly eccentric behaviours. Instead, focusing on the theme of display, I take a reception-oriented approach to Waterton's eccentricity, asking how Waterton's home, his collection and, indeed, the 'eccentric' naturalist himself were perceived by those who visited him, and how the published responses of those visitors functioned in the construction of Waterton's 'eccentric' reputation.

Histories of exhibitions, museums, gardens and country houses have not traditionally been primarily concerned with the experiences of the people who visited them. The existing literature on country houses is diverse – architectural historians have studied their design and construction, social historians have examined their domestic economies, historians of travel have shown how they attracted new types of visitor in the nineteenth century – yet rarely has this literature engaged directly with responses of people who actually went to these places.[9] Traditionally, studies of natural history collections have been concerned with where specimens came from, how they were obtained, how they were classified and catalogued, and perhaps how they were displayed. This picture has been greatly enriched by studies which have made visible the communication networks connecting collectors in all corners of the globe; nevertheless, such narratives still tend to tail off as each specimen finds its home in the museum under investigation. Recently, however, attention has begun to be directed to the cultural meanings which museum displays may have had for viewers in the past. One way of approaching this issue has been to consider how collectors, owners and curators intended their collections to be received.[10] This chapter, by contrast, examines actual recorded responses of visitors.

In the last chapter I used book reviews as indications of how Hawkins's books were received by readers; here I draw on published, first-hand visiting accounts in an attempt to recover something of the lived experience of real people as they encountered Waterton and his specimens in the flesh. The twelve accounts are of visits made between 1834 and 1865; they were pub-

Figure 4.2. The Nondescript. Image provided courtesy of Wakefield Cultural Services.

lished between 1834 and 1898.[11] One of the advantages of visiting accounts is that, by enabling audience differentiation right down to the level of the individual, they can reveal diversity of experience. Visitors to Walton Hall came from different social backgrounds, had different interests, and visited within different traditions; accordingly their experiences of Waterton's 'eccentricities' varied, and I highlight some of these differences along the way. Yet visiting accounts can also reveal commonalties of experience. Like book reviews, indeed like all literary productions, nineteenth-century visiting accounts were shaped by generic constraints. While this means they cannot be read simply as 'objective' accounts of personal experience, it also means they can be made to reveal shared conventions and expectations which governed how visiting was generally conducted, and how visiting experiences were interpreted, in the period. Often, it was against such conventions and expectations that eccentricity stood out.

The chapter begins by locating Waterton's visitors with respect to nineteenth-century traditions of country-house visiting; the responses of Waterton's visitors, including their comments on his 'eccentricities', can only be properly understood in this context. The second section of the chapter focuses on visitors' responses to Waterton's taxidermy. Waterton's idiosyncratic taxidermic method was considered eccentric by some contemporaries because it was different to that in common use. Moreover, Waterton's specimens themselves were felt to be eccentric because, especially in the case of his hybrid creations, they appeared to threaten fundamental classificatory boundaries. The third section considers the influence of Waterton's *Wanderings in South America* upon visitors' reception of the specimens. There was a unique relationship between the specimens and the adventurous stories in *Wanderings in South America*, and I show how visitors drew on these well-known stories in interpreting the contents of the museum. In the final section, locating Waterton in the context of the rise of celebrity culture in the nineteenth century, I consider visitors' responses to the celebrated naturalist himself. I explore how visitors made sense of their visits by drawing together their experiences of visiting Waterton's home, viewing his specimens, reading his stories and encountering him face to face. In recording and publishing their experiences in visiting accounts, visitors helped to shape Waterton's public identity; they were also instrumental in the construction of his posthumous reputation as one of England's great eccentrics.

Visiting Walton Hall: Expectations and Surprises

It was no wonder George Harley slept well. The renowned London physician had suffered no fewer than six surprises, four astonishments, a startlement, a bewilderment, a dismaying and one disappointment, all in less than twenty-four hours.[12] The disappointment had come first. Approaching the gates to Walton Hall park, one morning in the spring of 1856, Harley had glimpsed the Hall itself. There, on a small island near the edge of the vast lake, stood a modern, plain, three-storied house 'without any artistic features'.[13] Harley had been expecting to see the old feudal hall of the ancient Waterton family, with its drawbridge and stately oaken doors, but hardly a trace remained. In 1767, fifteen years before Waterton was born, the old hall had been demolished and replaced with the neo-classical building depicted in the engraving by Edwin Jones in Figure 4.3.[14] The height of fashion in its day, the Hall now grated with the current vogue for architecture in the gothic style. It was apt to disappoint those with a taste for visiting old buildings, an activity which had become popular of late as a result of a heightened interest in 'Olden Time' history – the history of the Tudors and Stuarts explored by Walter Scott in the Waverley novels, by Joseph Nash in his *Mansions of England in the Olden Time* (1830), and by their multifarious imitators in the periodical press.[15] The only old architectural relic which remained was the ruined watergate, which can be made out in Figure 4.3, just in front of the house itself. The Reverend J. Brooke, who visited Waterton in 1861, agreed with Harley that '[t]he mansion itself, although a spacious one, is the least interesting object in the whole domain, being quite a comparatively modern stone edifice in the Grecian style'.[16] An anonymous writer for the *Leisure Hour*, a cheap, weekly family journal of instruction and recreation produced by the Religious Tract Society, was similarly disappointed upon visiting Walton Hall that '[t]he mansion was not turretted, nor Elizabethan, nor of any style of the antique'.[17]

Visitors arrived at Walton Hall with expectations. In the last chapter, I argued that readers labelled books 'eccentric' when they failed to meet their initial expectations of what the book was going to be like. Like readers of Hawkins's fossil books, visitors to Walton Hall frequently expressed surprise that Walton Hall, with all it contained, was not how they had thought it would be. The unexpected was central to visitors' experiences of Waterton and his home. Indeed, paradoxically, on account of Waterton's increasingly widespread reputation as an 'eccentric' individual, experiences of the unexpected came ultimately to be part of the definition of a satisfactory visit.

I shall briefly outline the main traditions – the visit by personal invitation, the 'tour', the 'holiday' or day-long excursion – within which visiting tended to be conducted in this period. These constituted the overarching frameworks within which visitors encountered and interpreted Waterton, his home and his collection of taxidermic specimens. Country-house visiting was governed by a combination of explicit rules and tacit conventions, which Waterton expected visitors to abide by. And yet Waterton regularly astonished his visitors by flagrantly disregarding the conventions which were meant to govern the gentlemanly behaviour of country-house owners.

Harley arrived at the gates of Walton Hall on a bright and sunny morning. 'In response to the coachman's summons', he later recalled, 'a middle-aged woman appeared, and after scanning both me and my portmanteau, asked if I were Dr. George Harley, and being assured of this, she "dropped a curtsy", and said that

Figure 4.3. Walton Hall, engraved by Waterton's friend Edwin Jones. Holidaymakers enjoyed strolling in the grounds and fishing in the lake. From Waterton, *Essays on Natural History, Chiefly Ornithology*, 5th edn (London: Longman, 1844), frontispiece. Reproduced by kind permission of the Whipple Library, University of Cambridge.

the Squire was expecting me.'[18] Nineteenth-century visitors such as Harley were keenly aware of the processes by which they gained access to country houses, and they routinely recorded their experiences of planning, travel and admission in their visiting accounts. Some of the conventions governing access to Walton Hall had long been established within the context of gentlemanly professional or social visiting; others had been put into place more recently to cope with the growing numbers of 'tourists' and 'excursionists' who visited the Hall each year. Admission procedures varied, therefore, according to the purpose of the visit and the social standing of the visitor. Harley had received a personal invitation from Waterton to visit him in connection with their shared interest in the possible medical benefits of 'wourali', a compound which South American Indians used to poison arrows. Waterton had procured a quantity of the substance some decades earlier in the course of his first natural-historical expedition in British Guiana.[19] In his 'Reminiscences of Charles Waterton', published in the *Selbourne Magazine* in 1889, Harley records his receipt of the invitation and his reactions to it: his initial ambivalence at the prospect of visiting a gentleman of dubious reputation, more of which anon; his deliberations over how such a visit would reflect upon his own character; his overwhelming desire to replenish his stock of wourali; and his eventual decision to go ahead with the visit.[20] He details his travel to the Hall and his reception by the gatekeeper. He tells us that, once through the gates, the carriage proceeded towards the Hall, where he found the door open and the butler waiting to receive him. He was shown into a ground-floor reception room where he amused himself with a telescope until Waterton arrived. The telescope is just visible behind the table in Figure 4.4, a photograph from the collection of Rose Busk, sister of another of Waterton's visitors, Julia Byrne.

Julia Byrne (1819–94), author and wife of William Pitt Byrne, proprietor of the *Morning Post*, was like Harley a personal guest at Walton Hall. As her visit was of a social rather than professional nature her reception was more familiar than Harley's. Byrne's invitation and her initial responses to it are recorded in her *Social Hours with Celebrities: Being the Third and Fourth Volumes of 'Gossip of the Century'* (1898), which was edited by her sister Rose. Though they had never previously met in person, in 1861, Waterton invited Byrne via a mutual friend to retire to Walton Hall to recover from a recent 'family bereavement', presumably the death of her husband. Byrne had heard about the premature death of Waterton's wife some decades before, and in her visiting account she recalls being comforted by the prospect of spending time with 'one who could so profoundly enter into my state of mind'.[21] She recounts her journey to Wakefield with her friend, her transfer to Walton via a carriage sent by Waterton, and her arrival at the Hall. The friends were not ushered inside by gatekeepers and butlers like Harley; instead, as they drove up to the front of the house, Waterton and his family came outside to greet them personally. Byrne was deeply touched. 'Anyone', she writes, 'who had witnessed the affectionate cordiality of my reception, as the friend who introduced me and I reached the portico, where Mr. Waterton and his two sisters-in-law, who lived with him, came out to meet us, would have thought we had been bosom-friends for years'.[22]

Harley and Byrne would both ultimately become close friends of Waterton, and while they were treated differently on arrival because their visits were undertaken for different purposes, they were both personal guests who moved in high social circles; their visiting arrangements were thus relatively informal insofar as they were governed by etiquette rather than regulation. As Harley noted rather disparagingly in his reminiscences, however, such flexible arrangements would never do for the 'many tourists and "globe-trotters"' who tended to 'treat the Hall and Park as a public show place', nor the 'large parties of holiday-makers from the surrounding districts' who were allowed to use the park for picnics.[23]

'Tourists' in this period were men and women from the leisured classes engaged in literally touring a particular region. From the 1790s the wars on the

Figure 4.4. The drawing room at Walton Hall. The image, one of a pair of stereoscopic photographs, is from the library of Rose Busk, sister of Julia Byrne. © The Natural History Museum, London.

Continent had occasioned a boom in domestic tourism by closing off much of Europe, and by stimulating a nationalistic enthusiasm for British culture.[24] After the wars ended in 1815 Continental travel resumed but tours of Britain remained popular. Tourists were generally admitted to country houses on the grounds of letters of recommendation signed by a person known to the owner, a convention recognized as a matter of course by Waterton, who 'courteously admitted, and often acted as guide to, all persons who came with proper credentials'.[25] One such hopeful was the Reverend Thomas Frognall Dibdin (1776–1847), Vicar of Exning in Cambridgeshire and self-confessed 'bibliomaniac'.[26] Dibdin had already published an account of a tour of France and Germany which he had made in his capacity as Librarian to the Earl of Spencer, and now in 1837 he wished to view Walton Hall as part of his *Bibliographical, Antiquarian and Picturesque Tour in the Northern Counties of England and in Scotland* (1838).[27] In his account Dibdin records his nerve-racking experience of the admission procedures. He and his daughter turned up on horseback, unannounced but with a personal recommendation from the Reverend Mr Sharp, Vicar of Wakefield. Dibdin had made some enquiries in Wakefield and had been horrified to hear that Waterton was abroad and that 'strict orders had been given that NO ONE should be admitted upon any plea or pretence whatever'.[28] The Vicar had managed to console him: 'I had absolutely despaired, if the Vicar had not fortified me on starting to use his name and to take no refusal', Dibdin wrote. But then, unfortunately, the gatekeeper turned out to be a member of that 'fierce and peculiar race of human beings, of the feminine class, who are stern beyond all softening'. She set the pair on tenterhooks once again, almost refusing to let them pass.[29] At length, however, justice prevailed: the Vicar's name was flaunted, Dibdin and his daughter insisted they were 'both very peaceable and honest and would touch nothing', the gatekeeper softened, and they were allowed in.[30]

Large-scale changes in transport during the first half of the nineteenth century resulted in a heightened awareness of travel processes, and this was reflected in tourists' accounts of their visits. Sir George Head (1782–1855), a retired Assistant Commissary-General and regular contributor to the *Quarterly Review*, visited Walton Hall in the summer of 1835. He had already published a tour made some years earlier of North America,[31] and his *Home Tour Through the Manufacturing Districts of England* (1836) was devoted to visiting sites associated with agriculture, manufacturing, mining and transport. Ordinarily, Head was 'not partial to what is usually termed a "show place"', but the promise of a fine natural history collection tempted him to make an exception. Learning from a Wakefield innkeeper that 'full permission was granted to those who applied', while 'arrangements were at the same time made to protect the family from interruption', Head hired a horse and made his way to Walton Hall.[32] In his account he records in minute detail the circumstances of the journey: the mode

of transport, the route, the state of the roads and tracks at each point of the journey, the directions he received from local people, the gates he used, the bridge by
which he crossed the Barnsley Canal, and so on. Indeed, throughout the *Home
Tour*, Head discusses the advantages and disadvantages of canals, roads, railways
and so on, as he encounters them, a tendency common among visitors in this
period as they found themselves confronted by an increasingly wide range of
transport options.[33]

The 'Transport Revolution' also had a more direct impact on patterns of
country-house visiting. Emerging early in the century, but blossoming in the
1840s and 1850s, a third kind of visit, the day-long 'holiday' or 'excursion', was
the primary medium through which most middle- and working-class visitors
gained access to and experienced Walton Hall. Improvements in transport were
perceived to have been a key factor in the birth and rapid proliferation of this
new breed of traveller.[34] In addition, the rise of the excursion owed much to
pressure from Whig and radical reformers. As we saw in Chapter 2, advocates
of reform supported the foundation of Mechanics' Institutes and the publication of cheap tracts of 'useful knowledge' because they believed that education,
especially scientific education, would 'improve' the masses, thus equipping them
for more equal participation in social and political life. Travel and outdoor
pursuits were encouraged for the same reasons. From the 1820s onwards, for
example, protagonists of the Early Closing Movement argued that holidays were
essential for allowing workers to pursue improving leisure pursuits, while from
1844, a series of Railway Acts ensured that affordable third-class rail tickets were
made available for the benefit of working-class travellers. Charles Knight, publisher and leading light of the Society for the Diffusion of Useful Knowledge,
explained the benefits in his *Excursion-train Companion* (1851):

> The EXCURSION TRAIN is one of our best public instructors. It is also one of the
> cheapest ... From the great manufacturing and commercial towns, excursion trains
> are constantly bearing the active and intelligent artisans, with their families, to some
> interesting locality, for a happy and rational holiday. The amount of pleasure and
> information thus derived, and of prejudice thus removed, cannot be estimated at too
> high a rate.[35]

Many country-house owners were willing to open up their parks and gardens
to excursionists for the sake of affording 'rational recreation' to the working
classes, and some even competed to provide the best facilities and attractions.
At Walton Hall, visitors were allowed to wander in the grounds and could apply
for permission to fish in the lake: both activities are depicted in Figure 4.3. On
festive occasions there would be music, dancing and picnicking. In the early
years, admission to the Hall was granted freely, but as numbers increased stricter
regulations were enforced. One visitor wrote in 1848 that as many as 17,000

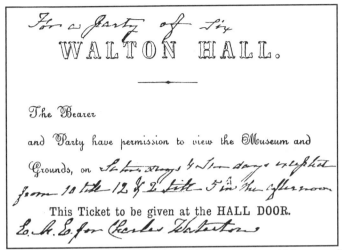

Figure 4.5. Waterton sent this ticket of admission to T. Allis, a Quaker and fellow naturalist, to pass on to a friend. It entitles a party of six to 'view the Museum and Grounds on Saturdays and Thursdays excepted from 10 till 12 & 2 till 5 in the afternoon'. Image provided courtesy of City of York Council, Local Studies Collection.

persons visited the Hall in one year but that 'owing to the misconduct ... of the many, that indiscriminate admission of all comers is now at an end'.[36] Visiting was restricted to certain days and times, and excursion parties were required to apply in advance for tickets, like the one shown in Figure 4.5.[37]

Visitors from different backgrounds visited Walton Hall within different traditions and their experiences varied accordingly. Nevertheless, visiting accounts reveal that certain overarching interpretive frameworks were applied by visitors from across the social spectrum in viewing the Hall, the grounds and the museum. Here I briefly discuss just one example, which will come up again later in connection with Waterton's *Wanderings in South America*. While visitors were often disappointed by the architecture of the mansion, they delighted in viewing and describing the grounds in accordance with the conventions of the Picturesque.[38] 'Once within the park gates', Harley recalled, 'a pleasing view met my eyes: it was a richly verdured and well-wooded glade sloping down to a large lake, the placid waters of which were glistening with the rays of the sun'.[39] The term 'picturesque' was originally coined simply to capture the notion that there were certain views, actual or ideal, which would look good in a picture. Here, Harley does not describe the park directly, in terms of its dimensions, say, or contents; rather, in the spirit of the picturesque, he describes it as a 'pleasing view' which met his eyes as he passed in his carriage through the gates of the hall. The reader might imagine the perfect vista emerging and then receding as the carriage continued down the drive, changing the aspect on the park with every creak of the wheels on the gravel. Harley attempts to capture this vista in words:

he paints an image of it in the mind's eye of the reader, evoking the texture of the landscape and the play of the light in his language.

Harley was by no means the only writer to experience the park as a series of views. Dibdin, for example, recalls the overwhelming delight he felt upon seeing the park from near the gateway. 'What a neighbourhood of wood and water!' he exclaimed. 'What a greensward – and what an embosomed mansion!'[40] Kinsey, by contrast, describes the park from a viewpoint in the interior of the Hall: 'The scenery of the extensive lake, as contemplated from the windows of the apartments looking to the east and south, is rich and beautiful.'[41] The writer for the *Leisure Hour* describes the park from a similar point of view:

> Far as the eye could range was the property of Waterton ... You looked around you on all sides, and you found a most tasteful mixture of wood and water, forest and sward, wild and cultivated; here nature running riot, and there, art adorning nature: and the whole was circumscribed within this valley, hills, feathered to the tops, bounding your view. There was excellent taste in selecting and decorating such a spot.[42]

Like Harley, this writer stresses the active role of the viewer – 'you look about you on all sides' – but his primary concern is to demonstrate his familiarity with the formal requirements of the picturesque, for the benefit and instruction of his self-improving readers. He observes the juxtaposition of 'wood and water, forest and sward', and judges it 'tasteful'. He satisfies himself that the park is a marriage of wild nature and carefully concealed art. He notices that the scene is perfectly framed, and in the words 'hills, feathered to the tops, bounding your view' he frames the reader's view too. His final seal of approval – 'There was excellent taste in selecting and decorating such a spot' – sums all all.

It was well known that wealthy landowners aimed to build their homes in locations which afforded good views of the surrounding land, often improving these views using fashionable landscape gardening techniques. Regular readers of the *Leisure Hour* would have been familiar with such practices from a recent series of articles on 'The Garden', in which the basic principles of picturesque landscape gardening, as well as the benefits to be derived from visiting such gardens, were laid out.[43] That Waterton had apparently followed fashionable practice and carefully combined wild nature and concealed art within his extensive estate meant that he could be judged a man not only of wealth but also of taste.[44] Waterton's taste was also evident from the decoration of the interior of the house. Harley, for example, wrote, 'no sooner had I crossed the threshold than all my feelings of disappointment vanished, and gave place to others of a totally different character as I found everywhere, in the furniture and the pictures on the walls, the usual signs of comfort and refinement of a country house'.[45] George Head described Waterton's drawing room as 'a room elegantly furnished; – besides some handsome pictures, with which the walls were ornamented, articles of *bijouterie* were tastily arranged on the tables; the general decorations well chosen, everything in its proper place, and the whole in first-rate aristocratic order'.[46] Against this background of respectable conformity, however, Waterton exhibited to his visitors surprising contrasts. Here was the crux of his perceived eccentricity.

Byrne, for example, was shown around Walton Hall by Waterton in person. Because she was an invited guest, rather than a tourist or an excursionist, she was allowed privileged access to areas of the house ordinarily closed to visitors. She was especially excited about being numbered among the 'few confidential friends' invited into Waterton's bedroom: 'He assured me I was one of the few strangers ever admitted', she explained. '"But then", he added, gracefully, "you are not, and never have been, a stranger to me; a common sorrow led us to understand and know each other from the first."'[47] The experience was 'not only interesting, but a surprise – indeed, a succession of surprises'.[48] The bedroom contained no bed – since the death of his wife Waterton had preferred to sleep on the floor – but it did contain tools and a menagerie of half-resurrected birds and animals, for it was here that Waterton prepared his taxidermic specimens. Harley was also admitted to the 'inner sanctum', and he was equally perturbed by his experience. Crossing the threshold, he was 'almost struck dumb with astonishment': there, right before his eyes, was a baboon, swinging in the air, suspended from the ceiling by strings.[49] Once he had recovered his composure, Harley gratefully allowed Waterton to initiate him in the mysteries of his art. Byrne, too, received a lesson in taxidermy, being allowed to assist in the transformation of the withered skin of a cock pheasant into an enchanting, lifelike specimen which she afterwards received as a gift.[50]

In the next section I examine responses to Waterton's taxidermy in more depth. Waterton rejected conventional taxidermic methods; he also rejected other conventions of institutionalized natural history, such as systematic nomenclature. Some fellow naturalists considered him eccentric on account of these transgressions. Non-specialist visitors, however, were more concerned with the appearances of the specimens themselves. They read them as eccentric objects, as challenges to the boundaries between life and death, between human and non-human, and between the different animal species.

Waterton's Taxidermy: Lifelike Wonders and Monstrous Hybrids

More surprises were in store for George Harley. Having recovered from his journey, he and Waterton embarked upon a tour of the 'museum'. Exotic mammals and fearsome reptiles guarded the entrance hall; gorgeous birds with iridescent plumage decorated the grand staircase, crammed into glass domes jutting out from the banisters; impressive paintings lined the walls from floor to ceiling, providing a magnificent backdrop to the collection. Richard Hobson called the display an 'unparalleled assemblage abounding in rarity, beauty and eccentricity'.[51] The photograph in Figure 4.6, also from Rose Busk's collection, shows the unusual arrangement of display cases on the staircase. Harley was overawed. Amazed that the specimens were 'so life-like in attitude and perfect in preservation', he muttered 'How splendidly they are stuffed.' The compliment was ill judged:

> [H]e suddenly wheeled round upon me, and with flashing eyes exclaimed – 'What do you mean? *Stuffed* did you say? Allow me to inform you that there are no stuffed animals in this house.'

Marching over to a nearby case Waterton extracted a perfectly preserved polecat specimen and presented it to the confused physician, ordering him, 'Take hold of the head, and hold it firmly':

> I did so, when he immediately gave the specimen a sudden jerky tug, and left the head in my hand. Astonished and dismayed, I immediately began to stammer out an apology. But instead of paying the slightest attention to what I said, he cut my speech short by saying, 'Look into the head – what do you see?' 'Nothing', was my answer ... No stuffing, no bones, no skull could I either see or feel. It was simply empty. It contained nothing but air ... Silently I gazed. Silently I wondered.[52]

Head in hand, Harley grappled with the significance of what he was witnessing. Waterton's specimens were completely hollow. Harley had never seen anything like it before.

Waterton prepared all of his specimens using a novel taxidermic method that he had devised himself. It has been argued that 'the collectors of the nineteenth century no longer regarded taxidermy as a problem but considered it a technique'.[53] While this was largely true of the growing class of professional naturalists, it was not true of Waterton, who in an essay of 1825 appended to *Wanderings in South America* complained:

> Were you to pay as much attention to birds as the sculptor does to the human frame, you would immediately see, on entering a museum, that the specimens are not well done. This remark will not be thought severe when you reflect that that which once was a bird has probably been stretched, stuffed, stiffened, and wired by the hand of a common clown.[54]

The standard taxidermic procedure in this period was to use arsenical soap as a preservative, and a combination of wires, stuffing and wooden supports to restore the skin to its original form.[55] Waterton was highly critical of this technique, insisting in a published 'Ornithological Letter' to his arch-enemy, the self-made zoologist and illustrator William Swainson (1789–1855), that anyone using it would find his preparations to be 'out of all shape and proportion'. 'If they should consist of birds', he wrote, 'they will come out of his hands a decided

Figure 4.6. The arrangement of specimen cases on the bannisters was often remarked upon. One visitor called it 'the enchanted staircase'. The photograph is from the library of Rose Busk, sister of Julia Byrne. © The Natural History Museum, London.

deformity. If of quadrupeds, they will represent hideous spectres, without any one feature remaining similar to those which they possessed in life.'[56]

Waterton took issue with institutional zoology not only in connection with taxidermic practice, but also as regards the respective roles of systematization and field observation in natural history. Swainson, who adhered to William MacLeay's quinary system of classification (according to which animals were arranged within a set of nested circles, each of which consisted of five subsidiaries), labelled Waterton 'an amateur', guilty of neglecting 'all which could be truly beneficial to science' on account of his rejection of systematic nomenclature in favour of common names.[57] Waterton responded by ridiculing Swainson's 'fond conceit of circles' and accusing him of obtaining his zoological information from the 'closet', rather than from the boundless fields of nature: 'Let me tell you', he warned his adversary, 'that the admeasurement of ten thousand dried bird-skins, with a subsequent and vastly complicated theory on what you conceive you have drawn from the scientific operation of your compasses, will never raise your name to any permanent altitude.'[58] Consequently, in developing his new taxidermic method, Waterton ensured it would allow him to make full use of his experience of having observed species in the wild. He described his method in an appendix to *Wanderings in South America* and in many other of his publications.

In place of arsenical soap Waterton used corrosive sublimate, a compound of mercury, as a preservative. Dispensing with wires and wooden supports, he allowed his skins to dry and harden very slowly until they could stand their own weight. His finished specimens were therefore hollow, as Harley discovered. Waterton adjusted the proportions of his specimens daily as they dried, over many weeks or even months. In contrast to professional taxidermists who tried to 'effect by despatch what could only be done by a very slow process',[59] Waterton indulged each of his specimens with painstaking care and attention, which, as a gentleman of leisure, he could afford. Based on his experience of observing birds and animals in the wild, he aimed to attribute to his specimens naturalistic forms and expressions. While 'closet' naturalists such as Swainson required above all that their specimens be easy to prepare, store and handle, Waterton desired that visitors to his museum at Walton Hall should exclaim, 'that animal is alive!'.[60]

And they did. The naturalist Reverend John Wood (1827–89), who visited Waterton on several occasions and published his observations in the *Magazine of Natural History*, wrote: 'Numbers of parrots and parakeets are displayed in all the attitudes which those mercurial birds assume, spreading their beautiful wings for flight, climbing up the boughs with their hooked beaks, ruffling their feathers, and scolding each other lustily, and, in fact, wanting nothing but movement to seem gifted with life.'[61] Hobson agreed: 'You can scarcely conceive it possible that even *living* nature could surpass, nay could equal the simple dead representations here displayed, to such a state of perfection has art attained.'[62]

What constituted 'lifelikeness' in the eyes of Waterton's visitors cannot be assumed to be self-evident. Visitors' remarks cannot have simply been based upon 'objective' judgements of verisimilitude between specimens and living animals: after all, most visitors would have seen only a tiny fraction of the species in a living state. It is widely accepted that realism on canvas, on the stage, or in literature has always depended on the skilful application of historically specific representational conventions, and I suggest the same was true for taxidermy in this period. Waterton actively promoted an elaborate discourse of taxidermic realism which placed value on such qualities as 'attitude', something which, he insisted, could only be known by a naturalist who had observed the living specimen in the field.

In addition, the equation of 'more realistic' with 'better' cannot be taken for granted. As the historian Stephen Bann observes, in this period 'there is no inevitable link between the science of anatomy and the rhetoric of life-like recreation'.[63] William Swainson kept his specimens as unmounted skins because they were easier to prepare and handle that way: for Swainson, this was the better way to preserve specimens. Waterton, however, encouraged his visitors to value 'life-likeness', as he defined it, above all other qualities. Emphasizing the painstaking care and attention lavished on each specimen, he encouraged visitors to apply a set of viewing conventions which served ideologically to place value on the skills and resources afforded him as a gentleman field naturalist. As a practice, Waterton's taxidermic method was perceived to be eccentric by some of his zoological contemporaries because it was not suited to the needs of professional men of science. One reviewer of *Wanderings in South America* called it 'a new and curious *mélange* of wireless cotton, sublimate, and sentimentalism'.[64] However, I suggest that individual specimens also played a part, for Waterton's specimens were themselves read as eccentric objects.

In Chapter 1, I suggested that the label 'eccentric' was applied in the nineteenth century to phenomena which appeared to challenge fundamental boundaries. Waterton's specimens challenged the boundary between life and death. Stephen Bann has likened Waterton's revivifying activities to those of Mary Shelley's Victor Frankenstein in her gothic novel, *Frankenstein: Or the Modern Prometheus* (1818).[65] Indeed, like Thomas Hawkins, Waterton explicitly characterized himself as a Promethean figure: 'You must possess Promethean boldness', he advised his readers, 'and bring down fire and animation, as it were, into your preserved specimen.'[66] All taxidermic specimens challenge the life/death dichotomy to some extent: they are all fragments of dead animals chemically preserved against the ravages of time and patched together to at least suggest the species in its living state. Waterton's specimens, however, were extreme on account of the great lengths he went to in order to restore his animal skins to a 'lifelike' appearance. A handful of them were even more Frankenstinian than the rest, being composed of the parts of various different animals stitched together. John Wood explained: 'Frogs, toads, and lizards, are grotesquely transmuted into caricatures of the human form; extraneous joints, limbs, claws, and horns, sprout from unexpected places.'[67] An example, *John Bull and the National Debt*, is shown in Figure 4.7. John Bull, a humanoid porcupine with a tortoise's shell, is weighed down by money bags and surrounded by a retinue of spiny mutant lizards, toads and millipedes with Latinate names: *Diabolus bellicosus, Diabolus ambitiosus, Diabolus illudens, Diabolus caeruleus.*[68]

'What have we here? half monkey and half man!'[69] The Nondescript was perhaps the most challenging of all Waterton's taxidermic creations, provoking reactions from amusement to curiosity, from disapproval to disgust (see Figure 4.2). The origins of the specimen were deliberately shrouded in mystery. In *Wanderings*, Waterton implied that he had crafted the specimen from the hindquarters of a howler monkey, but he would never say explicitly what the creature, whose portrait featured as the frontispiece to *Wanderings*, really was. Several reviewers of *Wanderings* scolded the author for trifling with natural his-

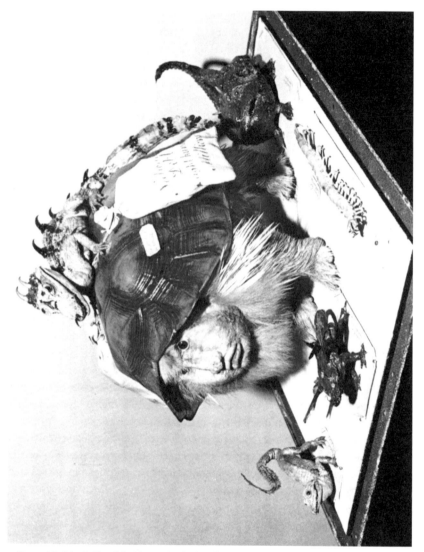

Figure 4.7. *John Bull and the National Debt*, by Charles Waterton. At £800 million the national debt, much of it accrued during the French wars, literally cripples 'Old Mr Bull', while he is simultaneously tormented by devils representing war, obsequiousness, jaundice, mockery, filth and gloom. Image provided courtesy of Wakefield Cultural Services.

tory. 'Such metamorphoses, instead of advancing, are prejudicial to, the science', wrote John Barrow in the *Quarterly Review*.[70] 'In this exhibition our author is surely abusing his stuffing talents, and laughing at the public', wrote Sydney Smith in the *Edinburgh Review*, before going on to suggest that the Nondescript was actually a representation of a Master in Chancery.[71] The *Literary Gazette* went one step further in sensationalizing the specimen, accusing Waterton of shooting 'a very respectable human native'.[72] A further reviewer, according to Wood, mistook the Nondescript for a portrait of the author, 'and thought that Mr. Waterton must be a very odd-looking person'.[73]

Harriet Ritvo has suggested that Waterton's Nondescript satirized 'the credulity implicit in the search for crowd-pleasing nondescripts'.[74] Whilst plausible as far as Waterton's intentions are concerned, from the point of view of the reception of the Nondescript, this argument overlooks the fact that tours and excursions to Walton Hall were often conducted *within* a crowd-pleasing tradition. During the seventeenth and eighteenth centuries the term 'nondescript' was used to designate a species which had not yet been described by naturalists. By the nineteenth century the word was more likely to be encountered not in learned zoological discourse but in advertisements for menageries and popular shows: in the late 1850s, for example, an American woman, Julia Pastrana, toured Europe promoting herself as the 'Bearded and Hairy Lady' or 'Nondescript'.[75] George Head wrote that of all the curiosities in the museum, this 'strangely human like being' was the one he viewed with most interest.[76] Wood likened the creature's face to that of 'a quaint and eccentric but genial-hearted old man'.[77] Harley thought it was 'a marvellous production, half nature, half art'. 'The face anyone might have admired', he wrote, 'but the body could only be loathed; it was all covered with hair, like the true body of a monkey, which in reality it was'.[78] Blurring the boundaries between the human and the non-human, the Nondescript had, like the hairy countenance of Julia Pastrana, the capacity to evoke sensations of disgust and nausea, yet any such sensations were invariably accompanied by an intense curiosity.[79]

Waterton's taxidermic specimens were read by visitors as eccentric objects in themselves. The Nondescript, the taxidermic hybrids and, to an extent, the naturalistic specimens too were viewed in a similar light to the 'freaks' and 'monsters' of travelling shows, and the human and non-human curiosities of eccentric biography, discussed in detail in Chapter 1. The hybrids, for example, challenged the boundaries between different animal species, rather like John Timbs's *Eccentricities of the Animal Creation*: the mermaid, the unicorn, the rhinoceros and the tree-climbing crab. Some visitors made the connection explicitly: 'the various odd compositions that meet the eye are made with a marvellous ingenuity', wrote Wood, 'that surpasses even the far-famed Japanese mermaids (of which, by the way, we have examined several), and bewilders the casual visitor to such an extent, that he is led to doubt whether the very staircase may not be a deception'.[80]

Visitors also found the specimens fascinating because they felt they expressed something of the 'eccentric' character of Waterton himself. Many of Waterton's composites, for example, were designed as religious caricatures. Throughout his life Waterton continued to feel aggrieved on account of the hardships suffered by himself and his ancestors as Roman Catholics. Even after the Catholic Relief Act of 1829, when penalties on Catholics were formally removed, he refused to take the oath of allegiance that was required for entry into public life. One way in which he expressed his grievances was through taxidermy. The most striking of the taxidermic caricatures, in the opinion of Brooke, was Waterton's representation of Philipp Melancthon, friend and collaborator of Martin Luther, 'for whom he had compounded an unsightly creature, made up of some large foreign lizard, furnished with sharp teeth taken from some carnivorous animal, and the claws of a cat'.[81] In conversation with Waterton, Brooke expressed surprise that he should deliberately offend the feelings of Protestant visitors with these 'repulsive figures, and the meaning they were intended to convey', to which Waterton replied simply that 'they none of them minded it much, but were highly amused'.[82] Henry Dixon concurred, writing that 'there was no lack of high-bred courtesy on [Waterton's] part to those of another creed'. 'The late Archbishop of Canterbury, who not unfrequently paid him a visit', he mused, 'must have smiled, as did many others, when he scaled that wonderful staircase with its pictured walls, and found on the two landings, among cases of humming birds, toucans, and the other results of his Wanderings, the "English Reformation zoologically illustrated".'[83]

A few visitors did, in fact, find the caricatures of Protestant leaders distasteful. Percy Fitzgerald, a school friend of Waterton's son, Edmund, recalled that these 'ingenious' creations 'were placed conspicuously in the hall, to the occasional scandal of visitors and sight-seers'.[84] One of those to be scandalized was Hobson, who accused Waterton of 'trenching on profanity'. 'If the subject be sacred', he wrote, quoting the Protestant dissenter Isaac Watts, 'all ludicrous terms, and jocose or comical airs should be excluded, lest young minds learn to trifle with the awful solemnities of religion.' Tellingly, however, he concluded that the hybrids were part and parcel with their creator's character: 'The Squire's inherent eccentricity of character, so universally known, from personal intercourse and from his various publications ... invariably allowed him, by sufferance, permission, or a sort of sanction, to say and do many things which the ordinary usages of society would not tolerate in others.'[85] A useful parallel can be drawn with William Martin. In Chapter 2 we saw that, through his 'eccentric' philosophical performances and publications, Martin was able openly to challenge, indeed abuse, Unitarians – something which would have been offensive and unacceptable if communicated in a more straightforward manner. Similarly, taxidermy apparently allowed Waterton to express his polemical opinions on religious matters to those who visited his museum, without causing as much offence as might have been the case had he announced his opinions straightforwardly in words.

Visitors felt they could obtain an insight into Waterton's 'eccentric' character through viewing his taxidermic hybrids. Many visitors were even more keen, however, to see the specimens that he had written about in *Wanderings*. This very widely read travel book was, I suggest, crucial to how many visitors interpreted Waterton's specimens, and so it is to *Wanderings* that I now turn.

Reading Stories, Reading Specimens

The publication in 1825 of *Wanderings in South America, the North-west of the United States and the Antilles, in the Years 1812, 1816, 1820 & 1824* can be located within the broad context of nineteenth-century scientific travel and exploration, which in turn produced a substantial body of scientific travel writing. Perhaps the best known works of the period to deal with South America are Alexander von Humboldt's and Aimé Bonpland's *Voyage aux régions équinoxiales du nouveau continent, fait dans les années 1799 à 1804*, published in Paris between 1805 and 1834 in thirty folio volumes, and Humboldt's *Personal Narrative of Travels to the Equinoctial Regions of the New Continent*, which followed in French and in English translation between 1814 and 1829.[86] While Humboldt and Bonpland gathered vast quantities of precise data regarding the geographical distribution of plant species in the course of their five-year expedition, Waterton, during his four short voyages, was concerned primarily with observing the habits of animal species and collecting specimens for his personal collection. The two projects were very different; nevertheless, Waterton's book can be considered as a contribution to the same, broadly defined genre.

As a genre, scientific travel writing has been analysed extensively in terms of its politics and its aesthetics.[87] Discourses of imperialism have figured prominently: during the eighteenth and nineteenth centuries, it is argued, 'travel writing became increasingly identified with the interests and preoccupations of those in European societies who wished to bring the non-European world into a position where it could be influenced, exploited or, in some cases, directly controlled'.[88] The practice of collecting objects, including natural history specimens, from far-flung places and accumulating them in European metropolitan centres has been characterized as a hegemonic form of possession of 'the other'.[89] In at least one respect, *Wanderings* is unusual in its *rejection* of imperialistic norms: rather than use the fashionable Linnaean nomenclature – Mary Louise Pratt argues convincingly that the Linnaean classificatory system played an important role in the construction of a Eurocentric 'planetary consciousness' – Waterton instead uses vernacular names.[90] In other respects, however, *Wanderings* does embody what Pratt has called an 'imperial stylistics'.

To give just one example, the following is from his first voyage up the Demerara river into the rainforests of Guiana:

> The finest park that England boasts falls far short of this delightful scene. There are
> about two thousand acres of grass, with here and there a clump of trees, and a few
> bushes and single trees scattered up and down by the hand of Nature. The ground is
> neither hilly nor level, but diversified with moderate rises and falls ... while the distant
> black rocks have the appearance of a herd at rest.[91]

Here Waterton domesticates the foreign landscape by explicitly comparing it to
an English park. This strategy was common in nineteenth-century travel writing, as James Duncan has shown in his analysis of British representations of the
Kandyan highlands of Ceylon, which in the 1840s were frequently compared
to the Lake District.[92] Waterton describes the landscape in typically picturesque
terms – it is a 'delightful scene'; the ground is 'diversified with moderate rises and
falls'; and so on – just as visitors to his home in Yorkshire described the grounds
at Walton Hall. *Wanderings* was a contribution to a well established genre of
travel writing, and would have been read as such.

Upon arrival at Walton Hall, visitors were handed a copy of *Wanderings* and
were encouraged to view the specimens in conjunction with the text. Case labels
gave page numbers of relevant passages. Head remarked favourably that the system provided a 'ready means of identifying with the object present its habits in
its native wilds'. One function of the labels, then, was to afford the curious visitor more detailed information on a particular specimen, and Head wrote that a
version of this system 'might be adopted in museums, and general collections of
objects of natural history, with advantage'.[93] Catalogues would point the viewer
to passages in the standard reference works, thereby increasing the collection's
instructive potential. Text was becoming an increasingly important aspect of
museum visits in this period, with the rise of the guidebook and the move to
fuller labelling of museum exhibits.[94] Yet at Walton Hall, the textual references
were especially pertinent. 'With regard to an exhibition such as the present', Head
reflected, 'wherein the owner's adventures are part and parcel with the creatures
exhibited, to refresh the memory by a recurrence to the narrative is doubly useful.'[95] At Walton Hall, the references pointed not simply to general texts, but to
observations and adventures relating to the very specimens on display.

Visitors viewing Waterton's cayman, for example, which was exhibited at the
top of the staircase, may have followed the instructions inscribed on a card displayed inside the glass case and turned to *Wanderings* to read about its origins.
During his travels in Guiana, Waterton had set about obtaining a perfect cayman
specimen for his museum. Trapping the beast was trouble enough, and Waterton
and some Indians who were assisting him spent several nights waiting in the forest before a cayman eventually swallowed the baited hook they had cast into the
Essequibo river. But how to kill it without damaging its skin? That was the really
awkward problem.

In his own account of the cayman capture, Waterton begins with a device commonly used in nineteenth-century travel writing: he sets the scene, building it up in layers like a theatrical tableau.[96] He firsts sets the mood, filling the background with distant menacing sounds: 'During the night, the jaguars roared and grumbled in the forest, as though the world was going wrong with them, and at intervals we could hear the distant cayman. The roaring of the jaguars was awful; but it was music to the dismal noise of these hideous and malicious reptiles.'[97] In the middle of the night, there was a tremendous shout. Arriving late on the scene, having misplaced his trousers, Waterton found his men fearfully contemplating a cayman, lashing about at the end of the rope. He lists his actors, and describes their characters, appearances, and costumes: 'We were, four South American savages, two negroes from Africa, a creole from Trinidad, and myself a white man from Yorkshire. In fact, a little tower of Babel group, in dress, no dress, address, and language.'[98] He positions his actors: 'I placed all the people at the end of the rope, and ordered them to pull till the cayman appeared on the surface of the water ... I now took the mast of the canoe in my hand ... and sunk down on one knee, about four yards from the water's edge, determining to thrust it down his throat.'[99] And the action begins:

> By this time the cayman was within two yards of me. I saw he was in a state of fear and perturbation; I instantly dropped the mast, sprang up, and jumped on his back, turning half round as I vaulted, so that I gained my seat with my face in a right position. I immediately seized his fore legs, and, by main force, twisted them on his back; thus they served me for a bridle.
>
> He now seemed to have recovered from his surprise, and probably fancying himself in hostile company, he began to plunge furiously, and lashed the sand with his long and powerful tail. I was out of reach of the strokes of it, by being near his head. He continued to plunge and strike, and made my seat very uncomfortable. It must have been a fine sight for an unoccupied spectator.
>
> The people roared out in triumph, and were so vociferous, that it was some time before they heard me tell them to pull me and my beast of burden farther inland.[100]

The action progresses only minutely before the actors are suspended in their positions: Waterton seated on the back of the cayman, the cayman plunging, the rope straining, the men roaring, and the whole entourage going nowhere. Waterton frames the adventure as a spectacle: he constructs himself a grandiose audience out of his 'people', who roar and cheer triumphantly at the moment of framing. Thus, the 'fine sight' is frozen, to be captured pictorially in the mind's eye of the reader, the 'unoccupied spectator'.

Not all visitors to Walton Hall may have taken the time to read extensively during their visits. Nevertheless the cayman adventure would have been familiar to many of them. *Wanderings* was reviewed in a wide range of periodicals, which frequently excerpted the cayman passage, as well as a number of hair-raising sto-

ries relating to snakes. Kinsey observed: 'The Quarterly and Edinburgh Reviews
have extended the knowledge of [Waterton's] most important works.'[101] The *Literary Gazette* excerpted the cayman story and three snake stories; Sydney Smith
excerpted the cayman- and two of the snake-wrestling adventures into the *Edinburgh Review*; the *Quarterly Review* excerpted two 'wondrous tough stories', one
cayman- and one snake-related; and by April 1826, the *British Critic* was declin-
ing to excerpt the passages in full, on the grounds that 'we believe they have
been transcribed into most of the newspapers'.[102] The story would also have been
familiar to visitors through its pictorial representations. At Walton Hall, Edwin
Jones's *Charles Waterton Capturing a Cayman*, was hung nearby the specimen
(Figure 4.1). A derivative engraving by the caricaturist and illustrator Robert
Cruikshank (brother of the more famous illustrator George Cruikshank), *It
Was the First and Last Time I Was Ever on a Cayman's Back*, was published in
January 1827 by G. Humphrey, St James's Street, and was republished in 1836
by Richard Nichols, a bookseller from Wakefield.[103] George Head reported that
this engraving, shown in Figure I.1, could 'be seen in many shop-windows'. 'Eve-
rybody is acquainted with the story of the crocodile', he wrote confidently.[104]

Visitors immediately recognized the cayman specimen as the one from
Wanderings. They described it as 'The fierce ill-looking cayman or crocodile, on
whose back Mr. Waterton fearlessly mounted, while his men were dragging the
monster of the deep from his native element'.[105] Head described his encounter
with the cayman like this: 'In a commanding position, with a lowering countenance, and an eye as horridly frowning as I ever beheld, stands extended at full
length the renowned crocodile, sufficient in his own person to recall to the mind
of the spectator that gallant equestrian feat which brought before the notice of
the world the latter part of his history.'[106] Other specimens, supplemented with
narratives from *Wanderings*, provoked similarly visceral responses. 'Yonder is a
Boa Constrictor', wrote Dibdin, 'coiled up to make his spring upon the unwary
traveller. His scales glisten and he moves along in splendid lubricity. I tremble to
approach him, and can hardly think I have passed him in safety.'[107] Dibdin first
sees the snake coiled and motionless, as if preparing to spring – so far it is only
he who is moving – but then the snake begins to move too, its scales catching the
light as it slithers along, terrorizing the 'unwary traveller'. In this word 'traveller',
we see the crux of Dibdin's fantasy; he is both the traveller on the staircase as he
motions towards the glass-cased specimen, and the traveller in the wilds of South
America as he happens upon a deadly serpent. The specimen reaches back to the
forests of Guiana to recall the place and the time of the original encounter and,
for a moment, Dibdin imagines himself in the picture with such vivacity that he
trembles at the thought.

Wanderings was crucial to how visitors interpreted the specimens on display
at Walton Hall, but it was also crucial to the construction of Waterton's personal

reputation as an eccentric character. In the final section I explore perceptions of Waterton's personal character further. On account of the popularity of Walton Hall as a visiting destination and the success of *Wanderings in South America*, Charles Waterton was a celebrity: 'His great fame, as an ornithologist', wrote Dibdin with characteristic enthusiasm, 'is not confined to his neighbourhood. It is spread all over Europe. It is acknowledged in the deepest wilds of American solitude.'[108] The rapid expansion of celebrity culture in the early nineteenth century had important repercussions for how 'the eccentric Waterton', and eccentricity in general, were understood.[109]

The Eccentric Naturalist as Specimen and Celebrity

It is generally agreed that something like our modern-day concept of celebrity came into being in the eighteenth century, having grown out of the ancient concept of 'fame': that is the character and qualities attributed to a person, usually in a good sense.[110] The reason, it is argued, that celebrity can have existed only since the eighteenth century is that celebrity is above all a media production. Thus, it cannot have been possible to conceive of celebrity until it was possible to conceive of a mass media, a condition which was only satisfied in the eighteenth century with the emergence of the public sphere. In the early eighteenth century, there was in fact a significant overlap between discourses of fame and eccentricity. In Chapter 1, for example, I cited the following poem by Aaron Hill, of 1743:

> Caesars, sometimes, and sometimes Marlbro's rise:
> Comets! That sweep new Tracks, and fright the Skies!
> Not to be measur'd, *These*, by War's *known* Laws:
> Form'd, for excentric Fame, and learn'd Applause!
> No *Gen'ral System* circumscribes their Ways.
> They move, un-rival'd: and were *born, to blaze!*[111]

Eccentricity, I argued, was a term used in association with lofty achievements and dazzling public successes. It was derived from an astronomical metaphor: individuals were called 'eccentric' because they were like comets. While in the nineteenth century discourses of eccentricity and fame would become more distinct from one another, at the level of language, at least, a connection remains to this day: the language of fame and celebrity is replete with astronomical symbolism.

The media explosion which occurred in the nineteenth century provided one of the conditions for an expansion – both in terms of scale and reach – of celebrity culture. Celebrity culture depends on individual reputations, or 'fame', but it is primarily constituted of representations – textual, oral, visual – of those individuals. News, pictures, advertisements and gossip are what make celebrity culture. In the early nineteenth century, a great number of people were afforded new opportunities for making, consuming and sharing these commodities. Leo Braudy, in *The Frenzy of Renown: Fame and its History*, has further linked the rise of celebrity to the emergence of the industrial age, which 'set the scene for individuals to make their way relatively unhampered by the traditions and restrictions of the past'.[112] Urban expansion, the extension of the political franchise and the revolutionary overthrow of monarchical authority created new opportunities, Braudy argues, for men of talent to build their own careers. In this context, 'acting and self-promotion abounded'.[113]

The rise of celebrity culture led in turn to a demand for celebrity face-to-face encounters. The possibilities for such encounters were multiplied by the circumstances, described at the beginning of this chapter, which for the first time made visiting the homes of famous people a realistic ambition for people from diverse social backgrounds. Scientific celebrities featured on the list of desired acquaintances alongside literary heroes, artistic geniuses and virtuoso performers.[114] Even the deceased were not quite exempt, as illustrated by William Hazlitt's 1826 essay, 'Of Persons One Would Wish to Have Seen'.[115] Set twenty years previously, the essay recalls a series of dinner-table discussions in which famous persons from the past were invoked and the merits of actually seeing them debated. Isaac Newton and John Locke are immediately dismissed because all that is really interesting about them is their books: they are 'not characters, you know ... what we want to see any one *bodily* for, is when there is something peculiar, striking in the individuals, more than we can learn from their writings, and yet are curious to know'.[116] The actor David Garrick, on the other hand, is somebody all would like to have seen in person on account of his legendary performative talents.[117] As Leo Braudy notes, Hazlitt's essay can be read as a variation on the ancient genre of 'dialogues of the dead', in which the greats of the past are revived and made to speak on topics of interest. Except there is a telling difference. Here, in the nineteenth century, the urge for dialogue, in which each great name argues his or her characteristic point of view, is replaced by the desire to see, to observe as in a theatre'.[118] With the living too, seeing 'as in a theatre' became central to the definition of a successful celebrity encounter.

The responses of Waterton's visitors show a familiarity with this new fashion for observing celebrities in their natural habitats, and a desire to interpret their own experiences in accordance with the conventions that were beginning to shape the celebrity visit. Harley's first encounter with Waterton, for example, was characteristically surprising. Having been shown into the drawing room of Walton Hall by a butler, he spotted the telescope on the table and decided to use it to survey the view from the window (see Figure 4.4). 'I had just placed my eye to the instrument', he recalled, 'when the door opened and I was face to face with my host – Charles Waterton, the famous traveller and naturalist.'[119] In this brief description of his first sighting of Waterton, Harley creates a sense of suspense (he is looking in the other direction as he waits), action (the door opens; Harley turns) and dramatic tension (the two men are suddenly 'face to face'). At the moment of recognition, Waterton's status as a celebrity figure is hammered home through repetition: he is Harley's host, he is 'Charles Waterton', he is 'the famous traveller and naturalist'. Harley's reminiscences were not published until the 1880s, by which time conventions for reporting celebrity encounters had become more firmly established. Visitors who published their accounts decades earlier, however, still attached considerable significance to meeting Waterton, and they explicitly remarked upon his celebrity status.

Waterton's physical appearance caused a great deal of excitement. His full-length portrait had appeared in the *Illustrated London News* in 1840, itself an indication of the celebrity status he had achieved (see Figure 4.8). Yet while this, along with written descriptions of his appearance, must have given visitors an idea of what to expect, many nevertheless expressed astonishment at Waterton's personal grooming and costume. While it was the fashion for men to wear their hair long, Waterton chose to wear his cropped short: Harley claimed the result reminded him of a man recently discharged from prison. His usual apparel, equally extraordinary, was described in the *Leisure Hour* thus:

> He was attired in a grey coat, very ill made, or at least ill fitting. Blue inexpressibles, with a vest of no particular cut, shoes thick and clumsy, stockings of blue, with a coloured neckerchief put carelessly round the throat, and a shocking bad hat, completed the habiliments. We thought at the time that the entire garniture, if put up for sale at a public auction, might probably have fetched about three and fourpence, including the seedy hat at one extremity, and the heavy shoes at the other. Yet we were standing in the presence of a man of ample fortune, cultivated taste, literary acquirements, and literary fame, owner of this charming retirement, and master of all he surveyed.[120]

Waterton's choice of footwear was a special source of consternation: 'He always wore short socks, – and *shoes* – the latter to enable him the better to climb rocks and trees, &c', wrote Brooke. Harley likewise observed that his 'tumble-down-at-heel slippers' were 'of the most unconventional kind'.[121] Waterton's costume was alarming not simply because it was unusual, but because it was felt to be inappropriate to his status as a gentleman: 'Although by no means rough and uncouth in manner or address, his clothing was plain and even coarse, such as might better have become his gamekeeper or groom', wrote Brooke. Hobson agreed that his 'personal apparel was of so peculiar a character, – of such a primitive style, and occasionally so much worn, and his hat generally in so dilapidated a condition, that he was now and then addressed by strangers as a person very much below his own grade in society'.[122] Exhibiting an awareness of fashionable practice in many areas of life – through his tasteful home furnishings, for example – Waterton was seen to challenge conventional distinctions of fashion, taste and, ultimately, social class by deliberately rejecting conventional dress in favour of practical clothing suited to outdoor work, tree-climbing and other ungentle-manly pursuits. Hobson insisted that Waterton's antipathy for fashion, and his insistence on wearing a 'shocking bad hat', was 'much to the regret of his most intimate friends'.[123]

Figure 4.8. Waterton generally wore a bright blue swallow-tail coat, and he always kept his hair very short, regardless of the fashions of the day. *Illustrated London News*, 24 August 1844, p. 124. Reproduced by kind permission of Cambridgeshire Libraries Archives and Information Services.

Waterton's alarming rejection of traditional class boundaries went further than his choice of apparel. 'He even carried his eccentricities so far as to occasionally pass himself off as one of his own servants, and he invariably had a hearty laugh when he related with what skill and success he performed the part', recalled Harley.[124] Several visitors recounted elaborate anecdotes substantiating this allegation. Brooke, for example, related the following story, apparently told to him by Waterton during his visit. Waterton was up in a tree looking at a bird's nest when a visiting party came by. Taking him to be the woodman, they began to question him about the owner. 'Well, he is a queer sort of an old chap', Waterton replied, 'and I advise you to be careful how you treat him, if you meet with him.' To add credibility to his performance, Waterton blew his nose loudly 'in an appropriate way' – appropriate to a woodman but not, of course, to a landed gentleman.[125] Julia Byrne wrote that 'pages might be filled with racy anecdotes of his eccentricities'; many of those she chose to include involved mistaken identity and comical reversals of status.[126] On one occasion, Waterton was weeding a path when a stranger walked past on his way to view the hall. Assuming he was the gardener, the stranger chatted to him for a while about Waterton's personality – 'They say he's a good-natured old fellow, although he's so queer … I'd almost rather see *him* than his museum.' Upon parting the stranger handed Waterton a coin – 'here's something to drink, for you look as if you wanted it'; Waterton thanked him kindly in his broadest Yorkshire accent. Having sent the stranger to the front door, he took a short cut to the back entrance and instructed the servants to invite the stranger to join him for dinner. In due course the clock chimed one o'clock and the butler solemnly announced 'Mr Tomkins', who was, of course, aghast at realizing his mistake.[127]

Predictably, perhaps, in an age increasingly preoccupied with celebrity, it was not uncommon to hear that, even when they did materialise, celebrity encounters could be disappointing. Stephen Gill, in his study of *Wordsworth and the Victorians*, observes this, citing John Ruskin's account of spotting Wordsworth in Rydal Chapel: 'We were rather disappointed in this gentleman's appearance especially as he seemed to be asleep the greater part of the time.'[128] Similar things were said of Waterton when it transpired that his appearance was closer to that of a common gardener than a man of wealth and fame. The *Leisure Hour* journalist reiterated this commonplace, advising his readers, 'If you would not break the charm which favourite living authors have flung around you, it may be as well, perhaps, not to pay them a visit.'[129] Yet while some visitors expressed disappointment that Waterton was not as grand as they had expected, they nevertheless took great delight in describing his oddities – of appearance, of dress and of manner – in intricate detail, and in composing elaborate anecdotes which they felt captured the eccentric essence of his personality. Indeed, one suspects that visitors would have been even more disappointed if they had found Waterton's dress and manners to be conventional. Hobson seems to have been of this opinion too. In his biography he criticized Waterton for pandering, especially in his later years, to the expectations of the masses: 'The supposed and ever-to-be-lamented imaginary proof of independence, by his usually sporting a somewhat eccentric and threadbare attire, pleased and captivated the vulgar, and by this *"aura popularis"*, "breeze of popularity", he was himself unfortunately caught, and permissively wafted on the wings of the multitude, at all events, for the moment.'[130] Writing in 1866, Hobson criticized Waterton for allowing himself to be swept up by the new celebrity culture: of pandering to the masses and wilfully displaying eccentric behaviours all for the sake of experiencing that momentary personal gratification which comes of being recognized as a celebrity.

Visitors wished to view Waterton because he was a celebrity. But I would suggest that they also had other motivations for wanting to set eyes on him that were closely related to Waterton's specimens and to his *Wanderings in South America*. In the years following the publication of *Wanderings*, Waterton's public reputation was slightly dubious. In particular, the truthfulness of his more extraordinary adventures was doubted. The *Quarterly Review* labelled these 'wondrous tough stories' tending to 'somewhat stagger our faith'.[131] The *Edinburgh Review* agreed that Waterton's 'stories draw largely sometimes on our faith', conceding that the book was all the more entertaining for this reason.[132] Visitors wanted to view Waterton, I suggest, because they felt that this would help them to verify the authenticity of the specimens and the stories about them in *Wanderings*.

The issue of credibility had gone hand in hand with travel writing and its criticism since the emergence of the genre. An early text often cited in this connection is the mid-fourteenth-century *Travels of Sir John Mandeville*.[133] Probably the most popular and most famous travel text of medieval Europe, Mandeville's accounts of the native peoples he encountered in central Asia, India and northern Africa were widely read and believed until at least the sixteenth century. However, the text was subsequently reclassified as a work of fiction, and Mandeville's *Travels* was transformed into a lesson on the untrustworthiness of foreign travellers and their tales. The following is an example the kinds of observations that Mandeville (about whom nothing more is known) was remembered for:

> There are many different kinds of people in these isles. In one, there is a race of great stature, like giants, foul and horrible to look at; they have one eye only, in the middle of their foreheads. They eat raw flesh and raw fish. In another part, there are ugly folk without heads, who have eyes in each shoulder; their mouths are round, like a horseshoe, in the middle of their chest. In yet another part there are headless men whose eyes and mouths are on their backs.[134]

The reliability of travel texts was still an issue when *Wanderings* appeared. The later decades of the eighteenth century had seen the publication of a number of highly successful, yet controversial travel texts of varying and sometimes ambiguous levels of fictionality. *Wanderings* would have been read within this context. Illustrating the point, reviewers explicitly compared Waterton's book to several well known titles.

Henry Southern, for example, explained in the *London Magazine* that he had been lead to expect *Wanderings* to be 'somewhat in the Munchausen vein'.[135] Rudolph Erich Raspe's anonymous *Surprising Adventures of Baron Munchausen* first appeared in London in 1785 and was reprinted many times. The book purported to be a truthful first-person account of a visit to Ceylon, composed only of 'REAL FACTS', but it was really a satirical attack on travellers and their incredible tales.[136] It begins with a story involving a crocodile, which bears a remarkable similarity to Waterton's adventure (see Figure 4.9).[137] About a fortnight into his stay in Ceylon, the story goes, Baron Munchausen accompanied some acquaintances on a shooting trip. Near the banks of a large piece of water, he was almost petrified at the sight of a lion approaching, apparently with the intention of devouring him. As if things couldn't get any worse, as he turned to flee, he encountered the voracious jaws of a crocodile waiting to receive him. The Baron collapsed to the ground in despair. But luck was on his side:

> After waiting in this prostrate situation a few seconds, I heard a violent but unusual noise, different from any sound that had ever before assailed my ears; after listening for some time, I ventured to raise my head and look round, when, to my unspeakable joy, I perceived that the lion had, by the eagerness with which he sprung at me, jumped forward as I fell, into the crocodile's mouth! the head of the one stuck in the throat of the other! and they were struggling to extricate themselves.[138]

Heroically, the Baron severed the lion's head with his hunting knife and, using the butt end, rammed the head down the crocodile's throat, thereby suffocating it. The preserved skin was placed on display at the public museum at Amsterdam, where, the narrator explains disapprovingly, 'the exhibitor relates the whole story to each spectator, with such additions as he thinks proper: some of his variations are rather extravagant'.[139]

A second traveller to whom Waterton was explicitly compared was the Scots explorer James Bruce. Bruce was not a fictional traveller like Baron Munchausen. However, his reports of his adventures in Abyssinia were initially considered highly suspect. In 1790, Bruce published *Travels to Discover the Source of the Nile in the Years 1768, 1769, 1770, 1771, 1772 and 1773*.[140] He had long since achieved notoriety, however, through relating his experiences in person. Upon his return from Abyssinia, where he claimed to have discovered the source of the Nile, Bruce was lionized. He was elected a fellow of the Royal Society, pre-

Figure 4.9. Waterton was compared to the fictional Baron von Munchausen, who cheated death twice when the lion that was chasing him leapt into the jaws of a nearby hungry crocodile. Illustration by Thomas Rowlandson from an 1811 edition of *Surprising Adventures of the Renowned Baron Munchausen*. Reproduced by permission of the British Library (RB.23.a.27647).

sented to the king, and courted by prominent members of the cultural elite, including the novelist Frances Burney and the politician, collector and writer Horace Walpole.[141] Select anecdotes were written up by James Boswell for the *London Magazine*. But then the tables turned. Crucial was Bruce's account of a Polyphemus banquet, the most remarkable features of which were, first, that the feast consisted mainly of raw steaks butchered from a live cow, which was left to bleed very slowly to death in the midst of the festivities and, second, that after eating the Abyssinians would pair off and copulate noisily in the dining room without causing remark or scandal.[142] This anecdote ultimately resulted in an unravelling of Bruce's credibility, which was not restored until years later. The comparison with Waterton was made in the *Literary Gazette* in 1826. Against the *Quarterly* and the *Edinburgh Reviews*, this reviewer asserted that 'we not only do not question his veracity, but (from the fine museum of natural history which he has collected in his extraordinary peregrinations, and other circumstances,) have reason for believing that his accounts are as accurate as those of Abyssinian Bruce, which were so cruelly doubted when his *Travels* were criticized'.[143] Waterton's collection was seen to be central to his credibility.

In later years, the periodicals tended to defend Waterton on the grounds of his 'character'. Whereas in the eighteenth century gentlemanliness was defined first and foremost by social rank, by the middle of the nineteenth century it had come to be redefined in terms of character, or 'moral order embodied in the individual', as Samuel Smiles put it in 1859.[144] The importance of character is exemplified in Harley's account of meeting Waterton for the first time. Having initially been concerned about Waterton's reputation as 'an unblushing and unblushable storyteller', Harley was reassured as soon as he set eyes on the naturalist.[145] 'No one could catch a sight of his beaming smile, or receive a glance from his speaking eye without feeling that no matter how bizarre might be the appearance of the outer man, the inner was lit up by a genial, highly cultivated, and sympathetic mind', Harley recorded. 'The cordial clasp he gave to my hand, and the words of warm welcome with which he greeted me, associated as they were with a winsome expression of truthful sincerity, at once drew my heart towards him.'[146] As a gentleman himself, Harley could perceive Waterton's virtue in every look, every gesture, every word. The clasp of his hand was cordiality itself, the glance of his eye spoke straight to Harley's heart, and Harley saw through Waterton's 'bizarre' exterior to the inner man.

In fact, the more 'bizarre' aspects of Waterton's character were themselves perceived as signs that *Wanderings* was a truthful work. Commentators defended Waterton not only on account of his gentlemanliness, but also on account of his supposed inherent eccentricity. In 1841, *Chambers's Edinburgh Journal*, comparing Waterton to yet another notorious traveller – Jonathan Swift's Lemuel Gulliver – explained:

> It was the fashion among the countrymen of Mr Waterton, when his 'Wanderings in South America' were first published in 1825, to laugh at the statements of the traveller, as being somewhat Gulliverian in their cast. But a more thorough acquaintance with the character of Mr Waterton has convinced the world of his being a man at once of talent and veracity, though with some oddities in his composition.[147]

Visitors to Walton Hall desired to witness these 'oddities' because they were an essential component of his character: 'the cayman-riding exploit is one so peculiar in its nature, and so characteristic of the eccentric naturalist', explained *Chambers's Edinburgh Journal*.[148] They desired to view the 'fierce ill-looking cayman ... on whose back Mr. Waterton fearlessly mounted', but they also desired to look upon the figure of Waterton himself as 'the first and only man who ever bestrode and rode a *cayman*'.[149] Only by meeting Waterton for themselves would they be able to confidently agree with the journalist who in 1852 wrote of the story, 'there is no doubt about its literal truth. It was a kind of feat which Waterton would glory in performing.'[150] Viewing Waterton, and witnessing his supposed eccentricities for themselves, was a means by which visitors could aspire to authenticate Waterton's stories and, indirectly, *their own* experiences at Walton Hall. It was for this reason that Thomas Dibdin and his daughter were devastated that Waterton was away from home when they visited. 'It was a sad disappointment to us both that such a man should be from home on the occasion of our visit', he wrote of his non-encounter – 'the master-spirit was wanting, to give pungency to anecdote and truth to conjecture. Were he only present to receive our bow and curtesy, it had been something.'[151]

In her *Social Hours with Celebrities: Being the Third and Fourth Volumes of 'Gossip of the Century'*, Julia Byrne related an anecdote which epitomized and satirized what, at least by the 1890s, was beginning to seem like a vulgar approach to celebrity, eccentricity and visiting. Throughout the nineteenth century, indeed, as the possibilities for visiting the homes of famous men and women extended to members of the middle and working classes, visitors in the older, polite traditions – visitors such as Byrne – began actively to define themselves in opposition to these new kinds of visitor, stressing their intimate, personal connections with their hosts, and smiling down on those who believed they could really know celebrated men and women through reading about them, and perhaps glimpsing them at the ends of hallways or riding out to distant parts of their estates. Such responses were part and parcel of a broader reaction against the commercialization and democratization of travel, which engendered a fashion for independent travel or 'antitourism' amongst those who perceived themselves as belonging to the cultural elite.[152] Byrne's anecdote goes as follows.

In the years following her first sojourn at Walton Hall, Byrne accompanied Waterton and his family on travels to the Continent. On one occasion in Aix-la-Chapelle the party was supposedly seated for dinner opposite a confused English tourist who, having identified Waterton's Yorkshire accent but being unaware to whom he was speaking, began to enquire about Walton Hall. After asking about the visiting restrictions he turned to the question of the Hall's owner, still unaware that he was talking to the man himself:

> 'I've heard he lives among a lot of wild beasts and birds and things. Is *that* true?'
> 'Yes, sir, it's quite true; but they're very well-mannered and quite harmless beasts, and so is he'.
> 'I'm glad of that, for I want to go there; and is the story about the crocodile true?'
> 'What story? What crocodile?'
> 'Why, that he rides all over the place on a crocodile; that's what I should like to see'.[153]

Byrne's anecdotal tourist learned everything he knew about Waterton from his books, and it was the Waterton from *Wanderings* – the celebrated traveller who rode a cayman in preference to more conventional modes of transport – that he hoped to encounter – or, more precisely, *to see* – at Walton Hall. He wanted something akin to the experience of the *Leisure Hour* journalist who, upon meeting Waterton, confirmed: 'Full of anecdote, he was a graphic describer, and we felt that in conversation we had the same man that appears in his books.'[154] This was the Waterton whom Edwin Jones began to depict in the unfinished watercolour shown in Figure 4.10. In a parody of a typically picturesque scene, Waterton and his cayman take a ride in Walton Hall park, while the Nondescript, Philipp Melancthon, John Bull and the National Debt and all the other taxidermic monsters from the collection look on, unconcerned. Visitors viewed the collection in the light of *Wanderings* and they did the same for Waterton himself. Drawing together their experiences of specimens, narratives and personal encounters, they viewed the celebrity naturalist as a specimen in his own 'eccentric' collection.

Figure 4.10. Charles Waterton, his home, his adventures in South America, and his taxidermic creations are all represented together in this unfinished watercolour by Edwin Jones. Reproduced by kind permission of the Governors of Stonyhurst College.

CONCLUSION

When Elizabeth Gaskell was starting to write her life of Charlotte Brontë in 1856, she copied into her manuscript a snippet from the *Quarterly Review*: 'Get as many anecdotes as possible. If you love your reader and want to be read, get anecdotes!'[1] In Chapter 1, I described how nineteenth-century authors of eccentric biography – many of them struggling hacks eager to compile new books from old material with a minimum of editorial input – took to the newspapers, scissors in hand, to collect snippets of text which they could use in their publications. Anecdotes were the perfect material for this endeavour. Discreet, circumscribed narratives, they could be cut and pasted at will. They could stand alone or alongside other anecdotes, but required little or nothing by way of contextualization, serving to evidence each character's alleged eccentricity through simple accumulation. As a unit of biographical narration, the anecdote is central to the history of science and eccentricity.

Anecdotes feature not only in eccentric biography, but also in many of the other types of source upon which this book has drawn: newspaper reports, autobiographical sketches, discovery accounts, travel narratives, book reviews, visiting accounts, reminiscences, and local histories. Anecdotes are perfect for sharing: readers of anecdotes can readily share in the writer's satisfactions, disappointments and surprises because anecdotes narrate events in terms not, necessarily, of what actually happened, but of what should have happened, what people might generally have agreed would have been most fitting for the occasion. For example, we don't, in reality, know whether Julia Byrne's anecdotal tourist, discussed at the end of the last chapter, ever existed, let alone whether he ever made it to Walton Hall to see Waterton ride around his estate on the back of a cayman. But we do know that stories *like* this have been told and retold about Martin, Hawkins and Waterton since their lifetimes, and constitute much of what we remember about them today. These anecdotal and seemingly rather trivial narratives are the building blocks of the stories we tell ourselves about science, eccentricity and our past.

By way of conclusion I wish first to make some general comments on the place of eccentricity within the history of early nineteenth-century science. I end with some speculative thoughts about the more recent historical legacies of Martin, Hawkins and Waterton and the significances that they hold for audiences today. In a sense this will take us full circle for, by and large, these are the received life stories which this study has aimed to build upon, challenge and reinterpret.

Natural science was by no means the only sphere of activity in which eccentricity was identified in the early nineteenth century. As we saw in Chapter 1, eccentric biography celebrated the lives of writers, painters, soldiers, actors, clerics, criminals, mothers, travellers, hermits: it appears that virtually every sphere of activity could be probed to reveal its eccentric characters. Nevertheless, I would suggest that natural knowledge was a particular focus for eccentricity. I have argued that objects, events and people were labelled eccentric if they were seen to transgress boundaries. Natural science, perhaps more than any other field of human activity, was in this period engaged in defining and negotiating new boundaries: between elite science and that intended for non-specialist audiences, for example; between genres of scientific writing; and between the various scientific disciplines. In this transformative period, boundary figures such as Martin, Hawkins or Waterton were peculiarly visible. Science, it turns out, is an excellent place to start if you want to observe the development of a discourse of eccentricity in this period. But, more than this, I would argue that eccentricity is a particularly revealing lens through which to observe those developments – popularization, specialization, professionalization – which are now so closely associated with the emergence modern science as we understand it.

The early nineteenth century witnessed a boom in what is commonly known as 'popular science'. Scientific books, articles, lectures and exhibitions aimed at general audiences flourished. Each of the case studies in this book has emphasised a different aspect of this complex trend. In Chapter 2, we saw how William Martin carved out a niche for himself within Newcastle's entertainment circuit as a performer: his distinctive blend of philosophical instruction, carnivalesque humour and aggressive self-promotion engaged and enlivened audiences in the pubs, theatres and streets of Newcastle, while his inflammatory publications provoked elaborate, performative responses from self-styled disciples and opponents. As a writer, Thomas Hawkins sought to endear himself to both specialist and non-specialist readers through an unconventional medley of anatomy, pastoral, visionary enthusiasm and self-aggrandizement. Charles Waterton captivated visitors to his museum through spectacular displays of creatures and curiosities, each with their own remarkable stories to tell. The eccentric reputations of Martin, Hawkins and Waterton depended on the existence of audiences for their various works and, crucially, on the willingness of members of those audiences to record their experiences for posterity. Potential audiences for science, and thus for scientific eccentricity, existed in the early nineteenth century as never before.

Specialization was discussed most explicitly in Chapter 3. Readers labelled Thomas Hawkins's fossil books 'eccentric' because they felt that Hawkins had disregarded the generic boundaries that were meant to distinguish the different kinds of literary work. Similar concerns were expressed in connection with the writings of Martin and Waterton too. Martin, as we have seen, was attacked by the radical newspaper editor William Mitchell for, amongst many other things, including poems in his system of philosophy. Waterton's *Wanderings* was also recognized as a generic hybrid because it intertwined scientific observation with personal narrative, poetic interludes and picturesque description. Indeed there are strong resonances between the comments of Hawkins's reviewers, discussed in detail in Chapter 3, and the comments of Waterton's critics. The *Monthly Review*, for example, characterized Waterton as 'merely an adventurous naturalist, full of an eccentric mixture of enthusiasm, sentiment, and lively humour'.[2] The *Dublin Review* wrote that 'Whatever comes from his pen is entertaining', but he 'does not profess science or method, nor does he much restrain himself within the rules of regular composition.'[3] The *Literary Gazette* remarked in familiar terms on the difficulties of reviewing such books:

> There is, as a frontispiece to this volume, 'a Nondescript;' and it is exceedingly appropriate, for the work itself is altogether nondescript ... The style is odd, the opinions odd, the sentiments odd, the descriptions odd; and, in short, the whole medley is odd, not even excepting the Natural History[4]

At least one critic explicitly noted a similarity between Waterton's and Hawkins' books, assuring readers that *Memoirs* 'may safely be compared with anything from the pen of Waterton – a kindred spirit ... with our saurian oryctologist'.[5] The comment is highly suggestive: though they came from very different backgrounds, were engaged in different branches of science, and wrote books aimed at different audiences, Waterton and Hawkins were evidently perceived to be, at some level, comparable. Writers of 'eccentric' scientific books cannot be dismissed as isolated, aberrant cases. Even in their own time they were afforded considerable critical attention and, perhaps, were also recognized as belonging to a type.

Taken together, then, the three case studies provide the beginnings of a foundation for a more general claim about eccentricity and the specialization of scientific writing in the early nineteenth century: as scientific publishing became more specialised, 'eccentric' hybrid works attracted notice and, increasingly, censure. Yet comparing the three cases also highlights the need for caution when dealing at this more general level. For example, generic hybridity did not cause Waterton anything like the trouble it caused Hawkins. Generic hybridity was not in itself anathema to travel writing in this period – indeed, with its inevitable intermingling of description, exposition and prescription, travel writing has been characterized by modern-day literary scholars as essentially a hybrid or 'androgynous' genre.[6] Waterton therefore enjoyed much greater freedom than Hawkins in this respect, something which becomes apparent only when the works of both are analysed closely within the appropriate contextual, generic frameworks.

Specialization was linked to eccentricity beyond the literary realm. Martin, Hawkins and Waterton each, in their own ways, worked at the margins of their respective fields, both institutionally and in terms of the content of their research. Martin concerned himself with phenomena which, though they may once have fallen within the remit of the natural philosopher, were now deemed to stand outside it – his attachment to perpetual motion is perhaps the most obvious case in point – and he vociferously denounced elitist scientific societies, most spectacularly the British Association for the Advancement of Science, which he nicknamed the British Association of Scientific Asses.[7] Waterton equally located himself in opposition to institutionalized zoology, outspokenly attacking 'professional' naturalists, such as the ornithologist William Swainson, for their attachment to 'closet' natural history, their lack of field experience, and their pointless obsession with nomenclature and systematization. Hawkins, by contrast, was desperate to be accepted into the geological elite. He relished his affiliation with the Geological Society of London and took every opportunity to strengthen his social and professional bonds with the leading lights of the discipline. Yet despite his efforts he was always kept at the margins: he was respected as an avid collector and restorer of fossils, but his knowledge of anatomy was found wanting. Indeed, in all three cases, a mastery of technical skill – inventing, fossil restoration, taxidermy – was considered as insufficient for recognition as a full member of the scientific elite, whether such recognition was sought or not.

Martin, Hawkins and Waterton were each considered eccentric in part *because* of their marginal status with respect to institutionally sanctioned science. Studying them, and the responses they provoked, can highlight the (sometimes explicit, sometimes tacit) boundaries which were taken to define what counted as legitimate topics for, and methods of, scientific inquiry. And yet studying them also shows that the picture was not as neat as it is sometimes represented. Any descriptive account of scientific culture in this period, including the move towards specialization, institutionalization and professionalization, needs to be able to make sense of the activities of figures like Martin, Hawkins and Waterton. It will not do simply to leave them out.

The complex relationship between science and religion, which crops up in each of the case studies, can serve as an example. While Waterton's taxidermic caricatures of key figures in Protestant history were meant as jokes, they nevertheless underlined a genuine hostility which Waterton, and other Roman Catholics, felt towards the Protestant establishment. Just one consequence of the established regime had been to restrict access of Catholics to a scientific career.

Examining eccentricity highlights boundaries, or barriers, which controlled who was allowed in and who was to be kept out of mainstream science. While the situation regarding Catholics was beginning to change in the period covered by this book, religion remained at the heart of debates over how science was to be defined. Martin and Hawkins were in many respects engaged in completely different enterprises from each other. They worked in different branches of science, they catered to different audiences, they held opposing views regarding the value of scholarship and learning, and they held very different attitudes towards orthodox, institutionalized science. And yet there are connections between them too. Both closely linked their scientific researches to sacred history and eschatology, and both styled themselves as prophets. When Hawkins published his second fossil book, *The Book of the Great Sea-Dragons*, in 1840, he had the frontispiece engraved by Martin's brother, John (see Figure C.1). John Martin's fierce, fiery, apocalyptic vision of the Earth ruled by ichthyosaurs, plesiosaurs and other extinct monsters is a world away from the serene landscape which Hawkins had designed himself six years previously as the frontispiece to *Memoirs* (see Figure 3.9). By 1840, Hawkins's own visions of this episode in Earth's history had become darker. Hawkins would not have aligned himself with the working-class millenarian tradition within which William Martin was writing. And yet this engraved image reminds us that, at least from a contemporary audience perspective, they were not living and working in separate worlds.

The vast historiography of science and religion tells us that, contrary to what might once have been assumed, there was not a straightforward rift between science and religion in the nineteenth century.[8] Nevertheless, the period covered by this book is generally understood to have been a crucial time for the negotiation of boundaries between the two. As we saw in Chapter 3, for example, this was the period in which 'scientific geologists' began actively to define themselves in opposition to 'scriptural geologists', the most extreme statement of this position being Charles Lyell's insistence that 'The physical part of geological inquiry ought to be conducted as if the Scriptures were not in existence.'[9] From responses to Hawkins, we might conclude that a discourse of eccentricity was one of the means employed by scientific geologists to *defend* their discipline: they marginalized Hawkins as an eccentric figure *because* he threatened these new boundaries. Perhaps more interestingly, however, by studying 'eccentric' characters like Hawkins, and Martin, we get to see the extent to which the contested relationship between religion and natural knowledge was still being fought out. Hawkins and Martin were right where it mattered most, in the middle of a highly politicized struggle over who had authority to talk about the place of humans and God in the history of the Universe.

Figures like Martin, Hawkins and Waterton tend to be left out of the historiography of science because they are seen as being anomalous. On the contrary, I

Figure C.1. Frontispiece to Thomas Hawkins's *The Book of the Great Sea-Dragons* (1840), engraved by William Martin's brother, John Martin. Reproduced by kind permission of the Syndics of Cambridge University Library.

would suggest that it is precisely because of their marginal, transgressive, threatening positions that they are potentially so useful to historians of science. On this basis, it is tempting to conclude with a call for greater recognition of eccentric characters in the history of science. The danger, of course, is that this would just lead us back where we began: to unwieldy accumulations of biographical curiosities and anecdotes, to a new kind of eccentric biography. This is surely to be avoided. When I have mentioned to people that I am writing about science and eccentricity, the most common response has been along the lines of, 'You must have plenty of material – weren't most of them eccentric in the nineteenth century?' The answer is no, in that only a relatively small fraction of likely candidates turn out to have been explicitly labelled and routinely characterized as eccentric figures by contemporary audiences. But, in a sense, it is also yes, in that there are so many 'likely candidates' to choose from in the first place. These are the nineteenth-century naturalists, collectors, inventors and natural philosophers whom *we think* are eccentric: because of their religious devotion, which perhaps we do not share; because of their beards, which would be deeply unfashionable if cultivated today; but, especially, because of their hearty adventures and curious anecdotes, which seem so alien to 'modern', professional science as we know it. To unthinkingly categorize these individuals as 'eccentric' on the basis of our own preconceptions about what science and scientists ought to be like would, of course, be unhelpfully anachronistic. And yet it is *not just* that people from the past appear eccentric to us because they are different. In academic history and popular memory alike, conceptions of nineteenth-century science – including science that is now thought of as 'mainstream' – are inextricably bound up with ideas about eccentricity, more so than for any other period. There is, I suggest, more work to be done in taking seriously (but not too seriously) the 'eccentric' anecdotes which make up so much of the history of science, including 'mainstream' science, in this period, as it has been told by the practitioners themselves, their audiences, their biographers and by historians.

On 27 May 1865, Charles Waterton died. He was buried in the grounds of Walton Hall in a spot he had chosen years before. The *Illustrated London News* described his funeral, illustrating the report with the engraving shown in Figure C.2. Describing in detail the small fleet of boats, draped in black, that ferried corpse, priests, family and friends to Waterton's final resting place between two oaks at the head of the lake, this illustrated report served to convey the extraordinary funeral to a wide and eager public audience.[10] That Waterton minutely planned the spectacle of his own death, and that he constructed his park as a mausoleum, suggests that through his natural-historical endeavours he may have been attempting to redeem, at the symbolic level, the ruins which history, since the Reformation, had left behind.[11]

Figure C.2. Once the coffin had been lowered into the vault on the far side of the lake, the ceremony terminated with the canticle 'Benedictus', and a linnet in one of the oaks overhead joined its song to the chanting of the clergy. *Illustrated London News* (17 June 1865). Reproduced by kind permission of Cambridgeshire Libraries, Archives and Information Services.

According to his friends, Waterton frequently reflected upon the ultimate fate of his home and his collection. Hobson claims Waterton was once reduced to tears as he speculated that 'the diversified assemblage of all his labours might be moonshine – a flash in the pan – a Will-o'-the'Wisp – a merely ephemeral gathering together of the productions of nature'. Having vented his grief, however, he suddenly smiled, and 'expressed a hope that the work of his hands might yet be a spark from which at some future period a flood of light, by numerously radiating powers, might illuminate another generation of the scientific of their day'.[12] While the family chapel housed his collection of sacred relics and his park was ultimately to house his own mortal remains, Waterton's museum housed the almost-living relics of his life's achievements. Like *memento mori*, the specimens pointed to the inevitability of death, yet, being perfectly preserved according to his unique taxidermic method, they would live on after their creator had perished. J. G. Wood, one of the most successful nineteenth-century popularizers of natural history and a friend of Waterton, speculated that 'the light and delicate fur and down are likely to outlast the edifice of stone and iron in which they are sheltered, and to be a more enduring memorial of their preserver than monuments of brass or cenotaphs of marble'.[13]

Walton Hall is still standing, though it is now known as the Best Western Waterton Park Hotel. Situated only 6.5 km from the M1 and only 10 km from the M62, it is, the website assures potential visitors, the ideal venue for weddings, conferences, business meetings and private functions.[14] The hotel is proud of its association with the Hall's former owner. In the Bridgewalk Restaurant there are pamphlets about Waterton and his adventures produced by the City of Wakefield Metropolitan District Council. Waterton's family name has even been integrated into that of the hotel (although this may be to avoid confusion with Walton Hall Stately Home and Country Club, a timeshare leisure resort in Warwickshire). While the parrots, snakes, baboons and the cayman are long gone, herons continue to nest undisturbed in the park and traces of Waterton's legacy remain.

Outside of the walls of the estate, Waterton's memory lives on in less insipid ways. To this day, *Wanderings* is acclaimed as a natural history classic, a source of inspiration for generations of naturalists – not least TV's David Bellamy and best-selling author Gerald Durrell. 'Of all the strange birds in the wilderness down in Demerara', writes Durrell, 'Charles Waterton was surely the strangest.' '[O]ne of the truly great eccentrics of all time', confirms Brian Edginton, his most recent biographer.[15] Indeed, Waterton has been framed throughout the twentieth century and into the twenty-first as one of the 'great' English eccentrics, appearing in countless volumes of modern eccentric biography.[16] Accounts of his life and adventures continue to feature regularly in Yorkshire newspapers and magazines – *Yorkshire Life*, the *Yorkshire Post*, *Yorkshire Ridings Magazine*, *The Dalesman* – and local guidebooks and collections of eccentric Yorkshire characters rarely leave him unnoticed.[17]

Waterton's natural history collection, complemented by maps, timelines, photographs, worksheets, video commentaries and other interpretive apparatus, is today the showpiece of Wakefield Museum: 'Explore the sights and sounds of the rain forest as you enter the world of Wakefield explorer and early conservationist Charles Waterton', entices the Museum's website; 'Follow his journey through South America and see the unique collection of rare species that he preserved for future generations.'[18] Indeed, alongside the continued recycling of age-old anecdotes about Waterton's personal eccentricities, there has recently been a move to reinvent him as 'the Father of wildlife conservation'.[19] The catalogue for an exhibition to celebrate the 200th anniversary of his birth, for example, characterizes him as prophetic figure: 'a man in advance of his time'.[20] A similar approach has been taken by Julia Blackburn in her 1989 biography, *Charles Waterton: 1782–1865, Traveller and Conservationist*. The jacket blurb explains, 'Since his death the memory of Waterton's personal eccentricities has flourished, while the originality of his ideas and work has often suffered. Using his surviving papers Julia Blackburn has redressed the balance in a biography that restores Waterton to his place as the first conservationist of the modern age.'[21]

That Waterton has been reinterpreted in this way is not trivial. The figure of
the eccentric naturalist as crackpot is entertaining up to a point, and such charac-
terizations of Waterton do indeed continue to entertain in a variety of contexts.
However, such figures cannot always readily be absorbed into the cultural fabric
of local communities in such a way as to shape, in a positive sense, how those
communities understand their relationship to the past. Eccentric characters can
make a difference only if they are appreciated for what was unique about them,
and if this uniqueness can be shown to have mattered. In Waterton's case, this has
been achieved by recasting him as an eccentric visionary. By reinventing Waterton as the first conservationist, Yorkshire magazines and heritage institutions
have transformed Wakefield into a place where something of not just local but
national, even global, importance happened. Nobody wants their local heritage
to be purely local in significance after all.

Martin and Hawkins have also been subjected to historical reinterpretation
and reassessment. In an effort to tie Martin's activities in more closely with commonly held ideas about the defining characteristics of his native region, local
commentators have tended to stress the successes of his inventing career. In
1895, in his introduction to *Men of Mark 'Twixt Tyne and Tweed*, a collection
of biographies originally serialized in the Newcastle *Evening Chronicle*, the antiquarian Richard Welford recorded the following reflections:

> In the North of England [a] spirit of self-reliance, leading to successful achievement,
> has received fruitful expansion, and here, between the banks of the Tyne and the
> shores of the Tweed, has found prolific development. Northumbria may not have
> given to the world eminent statesmen ...; nor learned divines ...; nor notable phi-
> losophers, beyond the harmless eccentric, William Martin, who styled himself the
> 'Philosophical Conqueror of Nations', and displayed his prolific profundities in
> twopenny tracts at the street corners; nor distinguished poets ...; but discoverers and
> inventors, leaders of industry and pioneers of commerce, workers and improvers in
> the wide fields of scientific research and mechanical construction, have been pro-
> duced here abundantly.[22]

Welford was a Martinian enthusiast. By 1895 he had amassed a collection of
148 of Martin's pamphlets and leaflets. Like other antiquarians of his day, he
preserved these documents because he perceived them to possess a special significance for his region's history.[23] For Welford, Martin embodied the essence
of Northumberland because he excelled, occasionally, at what Northumberland
did best – self-reliance, inventing, mechanical construction – whilst failing, by
conventional standards, at what it supposedly did worst – philosophy and other
abstract learned pursuits. Like Edward Walker and other of Martin's contemporary disciples, Welford framed Martin as a distinctive local character, to be
preserved against the homogenizing threats of modernity.

A continued emphasis on Martin's inventiveness can be traced through the twentieth century. Other aspects of his life, such as his millenarian prophetic views, barely even seem funny to the modern observer, let alone plausible, for there is no present-day context (even an evangelical religious one) within which they can be properly grasped. Mechanical ingenuity, however, continues to be appreciated. A page from *The Nor'-easter Staff Magazine of the North Eastern Electricity Board and Central Electricity Generating Board – North Eastern Region – Northern Division* for 1966 contains two relevant articles. The first, 'Awards to Outstanding Apprentices', shows two newly qualified electrical fitters proudly receiving gifts of books from Mr Coop, the Chief Engineer. The second is a short biography of Martin focusing on his successful inventions and, especially, his award of the Silver Medal by the Society of Arts.[24] A second example is a Newcastle *Evening Chronicle* article of 1972. The trigger for this article was the traffic nightmare resulting from emergency repair work to the new Scotswood Bridge – '"new" as opposed to the old one which worked'. Martin always maintained that he had pre-empted Robert Stephenson's design for the High Level Bridge over the Tyne, triumphantly opened in 1849, and the connection prompted columnist Eric Forster to give a sketch of Martin's life. Despite the implausibility of some of Martin's more ambitious schemes, Forster insists that 'beneath the outlandishness rested a certain genius'. 'To assume Willie was raving', he explains, 'is to deny him his place in local history.' If only he had been around in 1972, 'our Willie would have sorted out the Scotswood Bridge problems in no time at all'.[25]

Coupled, perhaps, to the rise of the heritage industry nationwide, the last twenty years or so have seen a variety of attempts to consolidate Martin's status as a figure of local importance. On 18 August 1987, for example, the *Evening Chronicle* announced the launch of a national appeal to find an unusual piece of headgear. It was hoped that the object would 'cap' an exhibition on the history of Wallsend, and anyone with information was urged to contact the newly founded Wallsend Heritage Centre by telephone. The headgear, explained Richard Ellam, the Heritage Centre's Manager, had once belonged to a local eccentric philosopher. He was concerned that it had found its way to one of the London museums, and wished to restore it to its rightful place in the north-east of England. The hat 'was made from an entire tortoise shell with a brass rim, and probably looked something like a second World War German army helmet'.[26] Now, an army helmet one might expect to see in a history exhibition, but a hat made out of a tortoise? The provenance of William Martin's tortoiseshell hat is a little confused. The earliest source to refer to it seems to be John Latimer's record of Martin's death in his *Local Records* of 1857, a standard reference work for local historians: 'His eccentricities in costume were not less remarkable: for some years previous to his death his head-dress consisted of the shell of a tortoise, mounted with brass.'[27] But descriptions vary: 'a helmet of cloth and metal, bearing a brass representation of the High Level Bridge, adorned with lions on each pier'; an 'extraordinary skull-cap decorated with tortoise-shell'.[28] Perhaps Martin had more than one hat – who knows. What is interesting is that by the late twentieth century this elaborate, enigmatic and elusive piece of costume dress had come at least in principle to represent a small town's historical identity.

The intense concern with the preservation and restitution of material culture which characterizes today's heritage industry has not been without its critics. Robert Hewison's *The Heritage Industry: Britain in a Climate of Decline* (1987) was written in response to the 'appalling' statistic that, at the time of writing, a new museum opened in Britain every week or so. While museums were commendable institutions individually, Hewison argued, 'collectively, their growth in numbers point[ed] to the imaginative death of this country'.[29] The efforts of the shortlived Wallsend Heritage Centre (it closed in 1993) to recover Martin's (possibly apocryphal) tortoiseshell hat would likely have been viewed by Hewison as yet further evidence that, in attempting to dispel fears that Britain was undergoing social and economic decay, the heritage industry could only resort to constructing 'fantasies of a world that never was'.[30] Against this trend for heritage baiting, however, historians such as David Lowenthal and Raphael Samuel have taken a more positive approach to understanding the roles of material relics, as well as a wide range of anachronistic practices, in shaping people's understanding of their relationship to the past.[31] I suggest that any future exploration of the relationship between science, eccentricity and historical consciousness might build upon this work. Rather than dismiss the activities of those seeking to remember nineteenth-century eccentric characters, it would, I believe, be more worthwhile to endeavour to understand these activities in their own terms, and to assess the positive contributions they may have made towards shaping the collective identities of local communities.

That Martin's hat has not, as far as I know, been recovered is perhaps fitting. For, in stark contrast to Waterton's stately museum and Hawkins's monumental collection, the products of Martin's scientific labours – his lectures, his demonstrations, his broadsides, his advertisements – were by their nature ephemeral. Although he probably intended his more substantial publications to endure through history, Martinian philosophical culture was primarily an oral culture, reliant on physical presence – he would 'visit all the principal Towns in the United Kingdom' – and verbal communication, either directly between himself and his audiences, or indirectly through the medium of his disciples. Oral culture continues to be an important aspect of Martin's historical legacy today.

On Tuesday 8 October 2002 a spectacle took place at the Queen's Hall, Hexham, Northumberland. Cult folk-rock band The Whisky Priests, joined by Northumbrian smallpipes virtuoso Chris Ormston, took to the stage to perform a new song about William Martin's Northumberland Eagle Mail:

If you ask it of me, 'Where have you been'
To report on the strangest beast I've ever seen
Up on the Town Moor, a travelling machine
Defied gravity without aid of legs or of steam.[32]

The song was performed as part of 'The Mad Martins', an evening of entertainment devised for the seventeenth Northumberland Traditional Music Festival (see Figure C.3). In addition to songs composed specially for the event, the

The Mad Martins

Thursday 17 July, 7.30pm
Basement Bar, City Screen, York

Keith Armstrong, Tyneside poet and narrator, with Gary Miller, vocals and guitar, and Glenn Miller, accordion - founders of the legendary folk-rock band The Whiskey Priests.

A cabaret-style performance of songs and poems reflecting the lives of the 'Mad' Martin family of Tynedale, including Jonathan Martin, the notorious incendiary of York Minster.

**Tickets £4 /£3 concessions
From City Screen Box Office
on 01904 541144**

◀ Keith Armstrong

Figure C.3. Promotional flyer for a performance of 'Mad Martins' in York in 2002 with 'Tyneside poet' Keith Armstrong. Reproduced by kind permission of Keith Armstrong.

performance featured dramatizations of the lives of Martin and his brothers, excerpts from William Bell Scott's reminiscences, and readings of Martin's poems and autobiographical writings. Local poet Keith Armstrong, coordinator of the Northern Voices creative writing project, scripted the event and acted as narrator. Armstrong cites Martin's independence, his eccentricity and his 'root-edness in Tynedale' as characteristics which have, since the 1970s, inspired him to base much of his poetry upon Martin's life, and to extend knowledge of Mar-tin's activities through community arts projects.[33] From 'The Lion of Wallsend' in 1975 to 'The Mad Martins' in 2002, Armstrong's work has helped to compen-sate for a disappointing loss of material relics by resurrecting an oral tradition of 'eccentric' Martinian performance.

Hawkins is best remembered today amongst the community of palaeontolo-gists.[34] His spectacular ichthyosaur and plesiosaur specimens are still scientifically important. The Sedgwick Museum of Geology in Cambridge and the Oxford University Museum of Natural History both display specimens donated by Hawkins in 1856 and 1874 respectively, and more than twenty of his specimens remain on display at the Natural History Museum in London, where they are amongst the most popular exhibits.[35]

His collecting endeavours and personal eccentricities are a favoured subject amongst palaeontologists engaged in negotiating the history of their discipline.[36] For example, Chris McGowan, Senior Curator of Palaeobiology at the Royal Ontario Museum, writes at length about Hawkins in the revealing titled *The Dragon Seekers: How an Extraordinary Circle of Fossilists Discovered the Dino-saurs and Paved the Way for Darwin*. 'Thomas Hawkins', notes Richard Forrest, of the Earth Sciences Department at New Walk Museum, Leicester, 'is one of the most bizarre characters in the history of Palaeontology – a science not nota-bly lacking in eccentrics.'[37] There remains, however, a degree of ambivalence as to Hawkins's suitability as a forebear of modern palaeontology. The Harvard palaeontologist Stephen Jay Gould, for example, introduces the seventh chap-ter of his *Finders, Keepers: Eight Collectors* with the equivocal pronouncement: 'In 1834, the eccentric and demented Thomas Hawkins extolled the virtues of untrammelled speculation.' Mike Taylor, Curator of Vertebrate Palaeontol-ogy for the National Museums of Scotland, writes in his *Oxford Dictionary of National Biography* entry that Hawkins was a 'social failure', arguing that this was 'presumably due to his eccentricity ... the underlying problem was evidently a serious personality disorder of unclear nature and onset'. Following Taylor, Gould ultimately concludes that 'we must face the tragedy of his probable mad-ness – for his eccentricities go far beyond the pleasurable English stereotype into the realm of tortured insanity'.[38] A world away from the 'worthy and eccentric man of genius' whom the comparative anatomist Richard Owen encountered in 1839, Hawkins has become a tragic lunatic – a man whose serious illness went untreated in an age before anybody recognized personality disorders or knew what to do about them.[39] For some at least, this means he must remain a marginal figure: an enthusiastic amateur collector rather than a founding father of palaeontological science.

Modern-day responses to nineteenth-century eccentric characters demonstrate a degree of nostalgia – nostalgia for a past in which members of the English landed gentry travelled and collected and set up 'nature reserves' in their estates just because they could; in which mechanically gifted labouring men invented things which worked; in which palaeontology, rather than being a technical branch of an esoteric scientific discipline, was simply a matter of strolling through the Somersetshire hills, hammer in hand, discovering enormous, breathtaking fossil monsters which would change the face of science and of history. Through such nostalgic discourse, eccentricity itself is made part of a broader heritage which pertains not just to how things were in Yorkshire, or Northumberland, or Somerset, but to how things were in general, in the good old days of popular memory, when science was for everybody, and Englishmen were at liberty to do as they liked and stand up for their beliefs, even when these beliefs were out of line with popular opinion. And yet, just as 'the pleasurable English stereotype' begins, perhaps, to seem just a bit too stereotypical, nineteenth-century eccentric men of science have begun to be systematically reclassified, as if to make them 'worthier' members of a common national and scientific heritage. Today, as in their own lifetimes, they are hybrid creations, compounded from autobiographical writings, portraits, anecdotal reminiscences, scientific publications and material relics which have been interpreted and reinterpreted over the best part of two centuries to suit a variety of present needs.

NOTES

Introduction

1. C. Waterton, *Wanderings in South America, the North-West of the United States and the Antilles, in the Years 1812, 1816, 1820 & 1824, with Original Instructions for the Perfect Preservation of Birds, Etc. for Cabinets of Natural History* (London: Mawman, 1825), pp. 227–31.
2. A. Desmond and J. Moore, *Darwin* (New York: Warner Books, 1992), p. 59.
3. M. Rudwick, *Scenes from Deep Time: Early Pictorial Representations of the Prehistoric World* (Chicago, IL, and London: University of Chicago Press, 1992), p. 142.
4. C. Babbage, *Passages from the Life of a Philosopher* (London: Longman, 1864), pp. 214–22.
5. Anon., *The Book of Wonderful Characters; Memoirs and Anecdotes of Remarkable and Eccentric Persons in All Ages and Countries. Chiefly from the Text of Henry Wilson and James Caulfield* (London: John Camden Hotten, *c.* 1869); D. Weeks and J. James, *Eccentrics* (London: Weidenfeld & Nicolson, 1995).
6. On eccentricity, see B. Cowlishaw, 'A Genealogy of Eccentricity' (PhD dissertation, University of Oklahoma, 1998); P. Langford, *Englishness Identified: Manners and Character 1650–1850* (Oxford: Oxford University Press, 2000), pp. 301–11; M. Gill, 'Eccentricity and the Cultural Imagination in Nineteenth-Century France' (PhD dissertation, University of Oxford, 2004); J. Gregory, '"Local Characters": Eccentricity and the North-East in the Nineteenth Century', *Northern History*, 42 (2005), 163–86; J. Gregory, '"Local Characters" and Local, Regional and National Identities in Nineteenth-Century England and Scotland', in A. Brown (ed.), *Historical Perspectives on Social Identities* (Cambridge: Cambridge Scholars Press, 2006), pp. 45–60; J. Gregory, 'Eccentric Lives: Character, Characters and Curiosities in Britain, *c.* 1760–1900', in E. Waltraud (ed.), *History of the Normal and Abnormal: Social and Cultural Histories of Norms and Normativity* (London and New York: Routledge, 2006).
7. On discipline formation, see, e.g., D. Cahan, 'Institutions and Communities', in D. Cahan (ed.), *From Natural Philosophy to the Sciences: Writing the History of Nineteenth-Century Science* (Chicago, IL, and London: University of Chicago Press, 2003), pp. 291–328. On the 'decline of science', see C. Babbage, *Reflections on the Decline of Science in England, and on Some of its Causes* (London: B. Fellowes, 1830); [David Brewster], Review of Babbage, *Decline of Science*, *Quarterly Review*, 43 (1830), pp. 305–42. On professionalization, see, e.g., P. Bowler and I. R. Morus, *Making Modern Science: A Historical Survey* (Chicago, IL, and London: Chicago University Press, 2005), pp. 329–37; S. F. Cannon, *Science and Culture: The Early Victorian Period* (New York: Dawson and Science His-

tory Publications, 1978), ch. 5; J. Morrell, 'Professionalisation', in R. C. Olby et al. (eds), *Companion to the History of Modern Science* (London and New York: Routledge, 1990), pp. 980–9; R. Porter, 'Gentlemen and Geology: The Emergence of a Scientific Career, 1660–1920', *Historical Journal*, 21: 4 (1978), pp. 809–36. On science in the Navy, see, e.g., R. Cock, 'Scientific Servicemen in the Royal Navy and the Professionalisation of Science, 1816–55', in D. Knight and M. Eddy (eds), *Science and Beliefs: From Natural Philosophy to Natural Science, 1700–1900* (Aldershot: Ashgate, 2005), pp. 95–111.

8. See, e.g., Bowler and Morus, *Making Modern Science*, ch. 16. On nineteenth-century audiences for science, and on the difficulties of defining 'popular science', see A. Fyfe and B. Lightman, 'Science in the Marketplace: An Introduction', in A. Fyfe and B. Lightman (eds), *Science in the Marketplace: Nineteenth-Century Sites and Experiences* (Chicago, IL: University of Chicago Press, 2007), pp. 1–19; see also the essays in that collection for specific examples of science popularization.

9. On mesmerism, see A. Winter, *Mezmerised: Powers of Mind in Victorian Britain* (Chicago, IL: University of Chicago Press, 1998). On phrenology, see, e.g., R. Cooter, *The Cultural Meaning of Popular Science: Phrenology and the Organisation of Consent in Nineteenth-Century Britain* (Cambridge: Cambridge University Press, 1984); J. v. Whye, *Phrenology and the Origins of Victorian Scientific Naturalism* (Aldershot: Ashgate, 2004). On psychical research, see, e.g., J. Oppenheim, *The Other World: Spiritualism and Psychical Research in England, 1850–1914* (Cambridge: Cambridge University Press, 1985).

10. On invisible technicians, see S. Shapin, *A Social History of Truth: Civility and Science in Seventeenth-Century England* (Chicago, IL: University of Chicago Press, 1994). On artisan collectors, see A. Secord, 'Science in the Pub: Artisan Botanists in Early Nineteenth-Century Lancashire', *History of Science*, 32 (1994), pp. 269–315. On women naturalists see, e.g., A. Shteir, *Cultivating Women, Cultivating Science: Flora's Daughters and Botany in England, 1760 to 1860* (Baltimore, MD: Johns Hopkins University Press, 1996).

11. B. Babcock, 'Introduction', in B. Babcock (ed.), *The Reversible World* (Ithaca, NY: Cornell University Press, 1978), pp. 13–33, at p. 14.

12. P. Stallybrass and A. White, *The Politics and Poetics of Transgression* (London: Menthuen, 1986), p. 17. In thinking about boundary transgression I have also been influenced by M. Douglas, *Purity and Danger: An Analysis of the Concepts of Pollution and Taboo* (1966; London: Routledge, 2002).

13. Babcock, 'Introduction', p. 32.

14. For an early 'diffusionist' model of science communication, relating specifically to the 'spread' of Western science, see G. Basalla, 'The Spread of Western Science', *Science*, 156 (1967), pp. 611–22. For alternative approaches to science communication, see, e.g., S. Friedman, S. Dunwoody and C. Rogers (eds), *Scientists and Journalists: Reporting Science as News* (New York: The Free Press, 1986); D. Miller, J. Kitzinger and P. Beharrell, *The Circuit of Mass Communication: Media Strategies, Representation and Audience Reception in the Aids Crisis* (London: Sage Publications, 1998); D. Nelkin, *Selling Science: How the Press Covers Science and Technology* (New York: W. H. Freeman & Co., 1987); R. Silverstone, *Framing Science: The Making of a BBC Documentary* (London: BFI Publishing, 1985).

15. See, e.g., M. Frasca-Spada and N. Jardine (eds), *Books and the Sciences in History* (Cambridge: Cambridge University Press, 2000); L. Henson, G. Cantor et al. (eds), *Culture and Science in the Nineteenth-Century Media* (Aldershot: Ashgate, 2004); J. A. Secord, *Victorian Sensation: The Extraordinary Publication, Reception, and Secret Authorship of Vestiges of the Natural History of Creation* (Chicago, IL, and London: University of Chicago Press, 2000); J. Topham, 'Scientific Publishing and the Reading of Science in Nineteenth-Cen-

tury Britain: A Historiographical Survey and Guide to Sources', *Studies in History and Philosophy of Science*, 31A (2000), pp. 559–612. Influential works in general book history include R. Altick, *The English Common Reader: A Social History of the Mass Reading Public, 1800–1900* (1957; Columbus: Ohio State University Press, 1998); R. Chartier, *The Order of Books* (Cambridge: Polity, 1994); R. Darnton, 'What Is the History of Books?' in Darnton (ed.), *The Kiss of Lamourette: Reflections in Cultural History* (London: Faber & Faber, 1990), ch. 1; E. Eisenstein, *The Printing Press as an Agent of Change: Communications and Cultural Transformations in Early-Modern Europe* (Cambridge: Cambridge University Press, 1979); A. Johns, *The Nature of the Book: Print and Knowledge in the Making* (Chicago, IL: University of Chicago Press, 1998).

16. 'Encoding and Decoding in Television Discourse', Stencilled Paper 7 (Birmingham: Centre for Contemporary Cultural Studies, 1973). An extract entitled 'Encoding/decoding' has been widely circulated in S. Hall, D. Hobson, A. Lowe and P. Willis (eds), *Culture, Media Language* (London: Hutchinson, 1980), pp. 128–62. On audiences and the media, see, e.g., I. Ang, *Desperately Seeking the Audience* (London: Routledge, 1991); D. Morley, *Television, Audiences and Cultural Studies* (London: Routledge, 1992); R. Silverstone, *Television and Everyday Life* (London: Routledge, 1994).

17. Results include W. Booth's 'implied reader', W. Booth, *The Rhetoric of Fiction* (Chicago, IL: University of Chicago Press, 1961); Gérard Genette's 'narratee', G. Genette, *Narrative Discourse*, trans. J. E. Lewin (Oxford: Blackwell, 1980); S. Fish's 'interpretative communities', S. Fish, *Is There a Text in This Class?: The Authority of Interpretive Communities* (Cambridge, MA: Harvard University Press, 1980); and H. Jauss's 'horizon of expectations' – his term for the set of cultural, ethical and literary expectations which form the basis upon which literary works are both produced and received – H. Jauss, *Towards an Aesthetics of Reception*, trans. T. Bahti (Minneapolis: University of Minnesota Press, 1982). For an overview of approaches to audience-oriented literary criticism, see, e.g., J. Raven, H. Small and N. Tadmor, 'Introduction', in Raven, Small and Tadmor (eds), *The Practice and Representation of Reading in England* (Cambridge: Cambridge University Press, 1996), pp. 1–21; S. Suleiman and I. Crossman, *The Reader in the Text: Essays on Audience and Interpretation* (Princeton, NJ: Princeton University Press, 1980); J. Tompkins (ed.), *Reader-Response Criticism: From Formalism to Post-Structuralism* (Baltimore, MD: John Hopkins University Press, 1980).

18. Secord, *Victorian Sensation*.

19. W. Martin, *A New System of Natural Philosophy, on the Principle of Perpetual Motion; with a Variety of Other Useful Discoveries. Patronised by His Grace the Duke of Northumberland* (Newcastle: Preston & Heaton, 1821).

20. T. Hawkins, *Memoirs of Ichthyosauri and Plesiosauri, Extinct Monsters of the Ancient Earth, with Twenty-Eight Plates Copied from Specimens in the Author's Collection of Fossil Organic Remains* (London: Relfe & Fletcher, 1834); T Hawkins, *The Book of the Great Sea-Dragons, Ichthyosauri and Plesiosauri,* גדלים חביבם*, Gedolim Taninim, of Moses; Extinct Monsters of the Ancient Earth; with Thirty Plates Copied from Skeletons in the Author's Collection of Fossil Organic Remains (Deposited in the British Museum)* (London: William Pickering, 1840); Anon., Review of Hawkins, 'The Book of the Great Sea-Dragons, Ichthisauri, and Pleriosuri [Sic.] Etc', *New Monthly Magazine*, 59 (1840), p. 431.

21. Waterton, *Wanderings in South America, the North-West of the United States and the Antilles*.

22. See, e.g., H. Roslynn, *From Faust to Strangelove: Representations of the Scientist in Western Literature* (Baltimore, MD: Johns Hopkins University Press, 1994).

23. E.g., on Darwin, see J. Browne, *Charles Darwin: The Power of Place* (New York: Alfred
    A. Knopf, 2002); on Newton, see P. Fara, 'Faces of Genius: Images of Isaac Newton
    in Eighteenth-Century England', in G. Cubitt and A. Warren (eds), *Heroic Reputations
    and Exemplary Lives* (Manchester: Manchester University Press, 2000), pp. 57–81; on
    Einstein, see A. Friedman and C. Donley, *Einstein as Myth and Muse* (Cambridge: Cam-
    bridge University Press, 1985); on John Hunter, see L. Jacyna, 'Images of John Hunter
    in the Nineteenth Century', *History of Science*, 21 (1985), pp. 85–108; on David Liv-
    ingstone, see J. MacKenzie, 'The Iconography of the Exemplary Life: The Case of David
    Livingstone', in Cubitt and Warren (eds), *Heroic Reputations and Exemplary Lives*, pp.
    84–104; on James Watt, see D. Miller, 'True Myths: James Watt's Kettle, His Condenser,
    and His Chemistry', *History of Science*, 42 (2004), pp. 333–60.

1 Defining Eccentricity in Early Nineteenth-Century Britain

1. Anon., *The Book of Wonderful Characters*, p. i.
2. J. S. Mill, *On Liberty* (1859; London: David Campbell, 1992), p. 64.
3. Anon., *The Book of Wonderful Characters*, p. ii.
4. Gregory, '"Local Characters" and Local, Regional and National Identities in Nineteenth-
   Century England and Scotland', p. 51.
5. Cowlishaw, 'A Genealogy of Eccentricity'; Langford, *Englishness Identified*, pp. 301–11;
   Gill, 'Eccentricity and the Cultural Imagination in Nineteenth-Century France'; Gregory, '"Local Characters": Eccentricity and the North-East in the Nineteenth Century';
   Gregory, '"Local Characters" and Local, Regional and National Identities in Nineteenth-
   Century England and Scotland'.
6. R. Recorde, *The Castle of Knowledge, Containing the Explication of the Sphere Both Celestiall and Materiall, Etc* (London, 1556), p. 247.
7. R. Savage, 'The Bastard', in Samuel Johnson (ed.), *The Works of Richard Savage* (London:
   T. Evans, 1777), pp. 91–5, at p. 91.
8. In 1665, for example, Samuel Pepys attended a lecture at Gresham College during which
   Robert Hooke argued that comets, such as the one of 1664–5, might return; Pepys
   recorded this as 'a very new opinion'; see S. Schaffer, 'Newton's Comets and the Transformation of Astrology', in P. Curry (ed.), *Astrology, Science and Society: Historical Essays*
   (Woodbridge: The Boydell Press, 1987), pp. 219–43, at p. 220. During the 1680s and
   1690s, Newton and Halley showed convincingly that comets moved in closed elliptical
   orbits that could be precisely described by natural laws.
9. J. Dryden, *The Conquest of Granada by the Spaniards* (London: Printed by T. N. for
   Henry Herringnan, 1672), p. 58.
10. Anon., *Shaftsbury's Farewel: Or, the New Association* (London: Printed by Walter Davis,
    1683), p. 1.
11. A. Hill, *The Fanciad: An Heroic Poem* (London: Printed for J. Osborn, 1743), p. 27.
12. W.N., 'Character of Robert Lecky Esq', *The Times* (26 June 1787), p. 1.
13. R. Williams, *Keywords: A Vocabulary of Culture and Society* (London: Croom Helm,
    1976), p. 13.
14. See, e.g., R. Altick, *Lives and Letters: A History of Literary Biography in England and
    America* (New York: Alfred A Knopf, 1966), p. 77 ff.
15. W. St Clair, *The Reading Nation in the Romantic Period* (Cambridge: Cambridge Uni-
    versity Press, 2004).

16. S. S. Genuth, *Comets, Popular Culture, and the Birth of Modern Cosmology* (Princeton, NJ: Princeton University Press, 1997), p. 3.

17. Schaffer, 'Newton's Comets and the Transformation of Astrology', pp. 234–5.

18. On prophecy, see, e.g., P. Curry, *A Confusion of Prophets: Victorian and Edwardian Astrology* (London: Collins & Brown, 1992); J. F. C. Harrison, *The Second Coming: Popular Millenarianism 1780–1850* (New Brunswick, NJ: Rutgers University Press, 1979). On almanacs, see, e.g., M. Perkins, *Visions of the Future: Almanacs, Time, and Cultural Change 1775–1870* (Oxford: Clarendon Press, 1996).

19. Mill originally published 'The Spirit of the Age' anonymously in *The Examiner* (nos. 77, 82, 92, 97, 103 and 107). The articles are reprinted in the *Collected works of John Stuart Mill, Vol. 22: Newspaper Writings by John Stuart Mill, December 1822–July 1831*, ed. A. P. Robson (London: Routledge, 1996), pp. 227–316 passim. The page numbers given here are from this reprint. On the spirit of the age, see J. Chandler, *England in 1819: The Politics of Literary Culture and the Case of Romantic Historicism* (Chicago, IL: University of Chicago Press, 1998), pp. 105–14.

20. [C. Robinson], 'Letter to Christopher North, Esquire, on the Spirit of the Age', *Blackwood's Edinburgh Magazine*, 28 (1830), p. 900.

21. Mill, 'Spirit of the Age', p. 228.

22. Ibid.

23. W. Hazlitt, *The Spirit of the Age: Or Contemporary Portraits* (1825; London: Oxford University Press, 1954), p. 85.

24. Ibid., pp. 85–86.

25. Ibid., p. 131.

26. Ibid., p. 132.

27. Ibid., p. 138.

28. Mill, 'Spirit of the Age', p. 230.

29. Ibid., p. 239.

30. Hazlitt, *The Spirit of the Age*, p. 1.

31. Ibid., p. 2.

32. Ibid., p. 3.

33. Mill, *On Liberty*, p. 63.

34. Ibid., p. 64. Mill's pronouncements have been cited in a variety of contexts in connection with nineteenth-century attitudes towards eccentricity; however their relationship to ideas about spirit of the age has not, as far as I know, been previously acknowledged. See, e.g., Cowlishaw, 'A Genealogy of Eccentricity' p. 129; Gill, 'Eccentricity and the Cultural Imagination in Nineteenth-Century France', p. 342; Langford, *Englishness Identified*, p. 309.

35. Altick, *Lives and Letters*, pp. 77–8. On prosopography in the nineteenth century, see, e.g., A. Booth, 'Men and Women of the Time: Victorian Prosopographies', in D. Amigoni (ed.), *Life Writing and Victorian Culture* (Aldershot: Ashgate, 2006), pp. 41–66.

36. W. Scott, 'Scott's Unfinished Autobiography', in R. Gibbons, compiler, *In Their Own Words: Autobiographical Writings of Seventeen of History's Greatest Thinkers and National Leaders* (New York: Random House, 1995), p. 335; Anon., 'Industrial Biography', *The Times* (28 December 1863), p. 5.

37. D. Higgins, *Romantic Genius and the Literary Magazine: Biography, Celebrity, Politics* (London: Routledge, 2005), p. 46.

38. L. Braudy, *The Frenzy of Renown: Fame and Its History* (Oxford: Oxford University Press, 1986), p. 13.

39. N. Hamilton, *Biography: A Brief History* (Cambridge, MA, and London: Harvard University Press, 2007), p. 10.
40. Altick, *Lives and Letters*, p. 82.
41. T. Carlyle, *On Heroes, Hero-Worship and the Heroic in History: Six Lectures. Reported with Emendations and Additions* (London: James Fraser, 1841), p. 47.
42. S. Collini, *Public Moralists: Political Thought and Intellectual Life in Britain 1850–1930* (Oxford: Clarendon Press, 1991), ch. 3. On character in science, see, e.g., A. Secord, 'Corresponding Interests: Artisans and Gentlemen in Nineteenth-Century Natural History', *British Journal of the History of Science*, 27 (1994), pp. 383–408, at p. 390.
43. S. Smiles, *Self Help: With Illustrations of Character and Conduct* (London, John Murray, 1859), p. 315.
44. On eccentric biography, with a focus on physical abnormality, see Gregory, 'Eccentric Lives', pp. 73–100.
45. Anon., *Eccentric Biography; or, Sketches of Remarkable Characters, Ancient and Modern, Including Potentates, Statesmen, Divines, Historians, Naval and Military Heroes, Philosophers, Lawyers, Impostors, Poets, Painters, Players, Dramatic Writers, Misers, &C. &C. &C. The Whole Alphabetically Arranged; and Forming a Pleasing Delineation of the Singularity, Whim, Folly, Caprice, &C. &C. Of the Human Mind. Ornamented with Portraits of the Most Singular Characters Noticed in the Work* (London: Vernor & Hood, 1801).
46. Anon., *Eccentric Biography: Or, Memoirs of Remarkable Female Characters, Ancient and Modern. Including Actresses, Adventurers, Authoresses, Fortune-Tellers, Gipsies, Dwarfs, Swindlers, Vagrants, and Others Who Have Distinguished Themselves by Their Chastity, Dissipation, Intrepidity, Learning, Abstinence, Credulity, &C. &C. Alphabetically Arranged. Forming a Pleasing Mirror of Reflection to the Female Mind. Ornamented with Portraits of the Most Singular Characters in the Work* (London: J. Cundee, 1803).
47. S. Johnson, Rambler, no. 60, 'On Biography', cited in R. Holmes, 'The Proper Study?', in P. France and W. St Clair (eds), *Mapping Lives: The Uses of Biography* (Oxford: Oxford University Press, 2002), pp. 7–18, at p. 10.
48. G. H. Wilson, *The Eccentric Mirror: Reflecting a Faithful and Interesting Delineation of Male and Female Characters, Ancient and Modern, Who Have Been Particularly Distinguished by Extraordinary Qualifications, Talents, and Propensities, Natural or Acquired, Comprehending Singular Instances of Longevity, Conformation, Bulk, Stature, Powers of Mind and of Body, Wonderful Exploits, Adventures, Habits, Propensities, Enterprising Pursuits, &C. &C. &C. With a Faithful Narration of Every Instance of Singularity Manifested in the Lives and Conduct of Characters Who Have Rendered Themselves Eminently Conspicuous by Their Eccentricities, the Whole Exhibiting an Interesting and Wonderful Display of Human Action in the Grand Theatre of the World. Collected and Re-Collected, from the Most Authentic Sources, by G. H. Wilson*, 4 vols (London: J. Cundee, 1806); Anon., *The Eccentric Magazine; or Lives and Portraits of Remarkable Characters* (London: G. Smeeton, 1812); R. S. Kirby, *Kirby's Wonderful and Eccentric Museum; Or, Magazine of Remarkable Characters. Including All the Curiosities of Nature and Art, from the Remotest Period to the Present Time, Drawn from Every Authentic Source. Illustrated with One Hundred and Twenty-Four Engravings. Chiefly Taken from Rare and Curious Prints or Original Drawings*, 6 vols (London: R. S. Kirby, 1803–20); Anon., *Eccentric Biography; Or Lives of Extraordinary Characters; Whether Remarkable for Their Splendid Talents, Singular Propensities, or Wonderful Adventures* (London: Thomas Tegg, 1826); F. Fairholt, *Remarkable and Eccentric Characters, with Numerous Illustrations* (London:

Richard Bentley, 1849); William Russell, *Eccentric Personages*, 2 vols (London: John Maxwell & Company, 1864); J. Timbs, *English Eccentrics and Eccentricities*, 2 vols (London: Richard Bentley, 1866).

49. Anon., *Eccentric Biography; Or Lives*, p. iii; Fairholt, *Remarkable and Eccentric Characters*, p. v.
50. On eighteenth-century biographical dictionaries, see, e.g., M. Frasca-Spada, 'The Many Lives of Eighteenth-Century Philosophy', *British Journal for the History of Philosophy*, 9 (2001), pp. 135–44. On scientific dictionaries see, e.g., R. Yeo, *Encyclopaedic Visions: Scientific Dictionaries and Enlightenment Culture* (Cambridge: Cambridge University Press, 2001).
51. J. Granger, *A Biographical History of England, from Egbert the Great to the Revolution: Consisting of Characters Disposed in Different Classes, and Adapted to a Methodical Catalogue of Engraved British Heads. Intended as an Essay Towards Reducing Our Biography to a System, and a Help to the Knowledge of Portraits*, 3 vols (London: T. Davies, 1769).
52. M. Pointon, *Hanging the Head: Portraiture and Social Formation in Eighteenth-Century England* (New Haven, CT, & London: Yale University Press, 1993), p. 54.
53. S. Sculptor, *Chalcographimania* (1813), cited in Pointon, *Hanging the Head*, p. 54.
54. Cited in Pointon, *Hanging the Head*, p. 63.
55. J. Caulfield, *Portraits, Memoirs, and Characters, of Remarkable Persons, from the Reign of Edward the Third, to the Revolution. Collected from the Most Authentic Accounts Extant. A New Edition, Completing the Twelfth Class of Granger's Biographical History of England; with Many Additional Rare Portraits*, 3 vols (1794; London: R. S. Kirby, 1813).
56. Pointon, *Hanging the Head*, p. 62.
57. J. Caulfield., *Blackguardiana; Or, a Dictionary of Rogues, Bawds, Pimps, Whores ... Illustrated with Eighteen Portraits of the Most Remarkable Professors in Every Species of Villany* (Bagshot: John Shepherd, 1795).
58. Anon., *The Eccentric Magazine*, p. v.
59. See, e.g., Chartier, *The Order of Books*; Gérard Genette, *Paratexts: Thresholds of Interpretation*, trans. J. E. Lewin (first published in French in 1987; Cambridge: Cambridge University Press, 1997).
60. Caulfield, *Portraits, Memoirs, and Characters, of Remarkable Persons*, preface.
61. Anon., *Eccentric Magazine*, p. v.
62. St Clair, *The Reading Nation in the Romantic Period*.
63. Ibid., pp. 110 ff.
64. Ibid., p. 205.
65. Ibid., p. 231.
66. Ibid., p. 118.
67. Ibid., pp. 229–30. See also, J. Secord, 'Scrapbook Science: Composite Carictures in Late Georgian England', in A. Shteir and B. Lightman (eds), *Figuring It Out: Science, Gender, and Visual Culture* (Hanover, NH: Dartmouth College Press, 2006), pp. 164–91.
68. C. Hulbert, *Breakfast of Scraps* (Shrewsbury: C. Hulbert, date unknown); Anon., *The Cabinet of Curiosities: Or Mirror of Entertainment. Being a Selection of Extraordinary Legends; Original and Singularly Curious Letters; Whimsical Inscriptions; Ludicrous Bills; Brilliant Bon Mots; Ingenious Calculations; Witty Petitions, and a Variety of Other Eccentric Matter* (London: Burkett & Plumpton, *c.* 1810); C. Hulbert, *The Select Museum of the World, or One Thousand Descriptions of Remarkable Antiquities, Curiosities, Beauties & Varieties of Nature & Art, in Asia, Africa, America & Europe* (Shrewsbury: C. Hulbert, 1822). On anthologies in the nineteenth century, see L. Price, *The Anthology and the Rise of the Novel: From Richardson to George Eliot* (Cambridge: Cambridge University Press, 2000). On the *Mirror of Literature*, see J. Topham, 'The *Mirror of Literature, Amuse-*

ment and Instruction and Cheap Miscellanies in Early Nineteenth-Century Britain' in G. Cantor, G. Dawson and G. Gooday (eds), *Science in the Nineteenth-Century Periodical: Reading the Magazine of Nature* (Cambridge: Cambridge University Press, 2004), pp. 37–66.

69. On Timbs's life and his career as a journalist, see J. Timbs, 'My Autobiography', *The Leisure Hour* (1871), passim.

70. H. Vizetelly, *Glances Back through Seventy Years: Autobiographical and Other Reminiscences*, 2 vols (London: Kegan Paul, Trench, Trübner & Co., 1893), vol. 1, p. 87.

71. Wilson, *The Eccentric Mirror*, p. iv; Kirby, *Kirby's Wonderful and Eccentric Museum*, vol. 1, p. iv; Fairholt, *Remarkable and Eccentric Characters*, preface; Timbs, *English Eccentrics and Eccentricities*, preface.

72. [John Jeaffreson], 'Stories of Inventors and Discoveries. By John Timbs', *Athenaeum*, 1675 (1859), p. 739; [John Doran], 'The Romance of London: Strange Stories, Scenes, and Remarkable Persons of the Great Town. By John Timbs', *Athenaeum*, 1966 (1865), pp. 13–14, at p. 14.

73. [Henry Fothergill Chorley], 'Remarkable and Eccentric Characters, with Numerous Illustrations', *Athenaeum*, 1111 (1849), p. 141.

74. [John Jeaffreson], 'English Eccentrics and Eccentricities. By John Timbs', *Athenaeum*, 2049 (1867), p. 155.

75. [John Doran], 'Things Not Generally Known. Curiosities of History; with New Lights. A Book for Old and Young. By John Timbs', *Athenaeum*, 1526 (1857), pp. 110–11.

76. [Frederick Stephens], 'Anecdote Biography. By John Timbs', *Athenaeum*, 1724 (1860), p. 627.

77. [Jeaffreson], 'English Eccentrics and Eccentricities', p. 155.

78. E. Sitwell, *The English Eccentrics* (London: Faber & Faber, 1933).

79. Reader's Digest, *Great British Eccentrics: They Entertained, Exasperated and Charmed a Nation* (London: The Reader's Digest Association, 1982); J. Keay, *Eccentric Travellers* (1982; London: John Murray, 2001); P. Bushell, *Great Eccentrics* (London: George Allen & Unwin, 1984); D. Dutton and G. Nown, *Oddballs! Astonishing Tales of the Great Eccentrics* (London: Arrow Books, 1984); M. Nicholas, *The World's Greatest Cranks & Crackpots* (London: Octopus Books Limited, 1984); J. Michell, *Eccentric Lives and Peculiar Notions* (London: Thames & Hudson, 1989); J. Timpson, *Timpson's English Eccentrics* (Norwich: Jarrold Publishing, 1991); C. Pickover, *Strange Brains and Genius: The Secret Lives of Eccentric Scientists and Madmen* (New York: Quill, 1999); K. Shaw, *The Mammoth Book of Oddballs and Eccentrics* (London: Robinson, 2000); J. John, *Eccentrics* (London: Duckworth, 2001).

80. A. Smith, *The Natural History of Stuck-Up People* (London, 1847); A. Reach, *The Natural History of Bores* (London: D. Bogue, 1847); J. Smith, *Vagabondiana; Or, Anecdotes of Mendicant Wanderers through the Streets of London; with Portraits of the Most Remarkable Drawn from Life* (London: Published for the proprietor, 1817). On Cries, see S. Shesgreen, *Images of the Outcast: The Urban Poor in the Cries of London* (Manchester: Manchester University Press, 2002). Single-leaf prints depicting hawkers and their calls were produced from the sixteenth century. 'Cries' first began to be produced as illustrated books around 1750, flourishing during the first half of the nineteenth century. Examples include T. Ticklecheek, *The Cries of London, Displaying the Manners, Customs & Characters, of Various People Who Traverse London Streets with Articles to Sell. To Which Is Added Some Pretty Poetry Applicable to Each Character. Intended to Amuse and Instruct All Good Children, with London and the Country Contrasted* (London: J.

Fairburn, 1797); Anon., *The New Cries of London, with Characteristic Engravings* (London: Printed and sold by Darton & Harven, 1804); and J. Smith, *The Cries of London: Exhibiting Several of the Itinerant Traders of Ancient and Modern Times. Copied from Rare Engravings, or Drawn from the Life, by John Thomas Smith, Late Keeper of the Prints in the British Museum. With a Memoir and Portrait of the Author* (London: John Bowyer Nichols & Son, 1839).

81. Fairholt, *Remarkable and Eccentric Characters*, pp. iv–v.
82. J. Timbs, *Eccentricities of the Animal Creation* (London: Seeley, Jackson & Halliday, 1869).
83. On monsters and natural history, see, e.g., H. Ritvo, *The Platypus and the Mermaid and Other Figments of the Classifying Imagination* (Cambridge, MA: Harvard University Press, 1997).
84. Kirby, *Kirby's Wonderful and Eccentric Museum*, vol. 6, pp. 488–9.
85. On the exhibition of human 'freaks' and 'monsters', see, e.g., R. Altick, *The Shows of London* (Cambridge, MA: The Belknap Press of Harvard University Press, 1978); B. Benedict, *Curiosity: A Cultural History of Early Modern Inquiry* (Chicago, IL, and London: University of Chicago Press, 2001); R. Bogdan, *Freak Show: Presenting Human Oddities for Amusement and Profit* (Chicago, IL, and London: University of Chicago Press, 1988); H. Deutsch and F. Nussbaum (eds), *'Defects': Engendering the Modern Body* (Ann Arbor: University of Michigan Press, 2000).
86. On Hudson, see Fairholt, *Remarkable and Eccentric Characters*, pp. 53–75; Anon., *The Eccentric Magazine*, vol. 1, pp. 15–18; Anon., *Eccentric Biography; Or, Sketches*, pp. 169–71; Anon., *Eccentric Biography or Lives*, pp. 130–1. On Lambert, see Anon., *The Eccentric Magazine*, vol. 2, pp. 241–8; Wilson, *The Eccentric Mirror*, vol. 1, no. 1, pp. 1–34; Kirby, *Kirby's Wonderful and Eccentric Museum*, vol. 2, pp. 408–10. On Topham, see Kirby, *Kirby's Wonderful and Eccentric Museum*, vol. 1, pp. 157–63; Anon., *Eccentric Biography; Or Lives*, pp. 40–5; Fairholt, *Remarkable and Eccentric Characters*, pp. 47–57. On Buchinger, see Kirby, *Kirby's Wonderful and Eccentric Museum*, vol. 2, pp. 1–2. On O'Brien, see Kirby, *Kirby's Wonderful and Eccentric Museum*, vol. 2, pp. 332–7; Wilson, *The Eccentric Mirror*, vol. 1, no. 2, pp. 26–34.
87. Kirby, *Kirby's Wonderful and Eccentric Museum*, vol. 3, p. 381.
88. Ibid., vol. 3, p. 381.
89. Timbs, *English Eccentrics and Eccentricities*, vol. 1, p. 97.
90. On Fuller, see Kirby, *Kirby's Wonderful and Eccentric Museum*, vol. 1, pp. 27–31. On Dancer, see Kirby, *Kirby's Wonderful and Eccentric Museum*, vol. 3, pp. 169–75; Anon., *Eccentric Biography; Or Lives*, pp. 145–51. On Elwes, see Kirby, *Kirby's Wonderful and Eccentric Museum*, vol. 3, pp. 258–76; Anon., *The Eccentric Magazine*, vol. 2, pp. 5–28; Anon., *Eccentric Biography; Or Lives*, pp. 49–69; Fairholt, *Remarkable and Eccentric Characters*, p. 130. On Jennings and Ostervald, see Anon., *Eccentric Biography; Or Lives*, pp. 27–9, pp. 139–40. On Ward, see Timbs, *English Eccentrics and Eccentricities*, vol. 1, pp. 79–91. On Taylor, see Wilson, *The Eccentric Mirror*, vol. 1, no. 2, pp. 34–8. On Cooke, see Timbs, *English Eccentrics and Eccentricities*, vol. 1, pp. 87–93; Kirby, *Kirby's Wonderful and Eccentric Museum*, vol. 5, pp. 297–329. On Welby, see Anon., *Eccentric Biography; Or Lives*, pp. 313–16; Kirby, *Kirby's Wonderful and Eccentric Museum*, vol. 5, pp. 43–5. On Bishop, see Kirby, *Kirby's Wonderful and Eccentric Museum*, vol. 3, pp. 131–5; Wilson, *The Eccentric Mirror*, vol. 1, no. 1, pp. 31–6. On Crabb, see Wilson, *The Eccentric Mirror*, vol. 1, no. 9, pp. 46–8; Anon., *The Eccentric Magazine*, vol. 1, pp. 97–100. On Bigg, see Kirby, *Kirby's Wonderful and Eccentric Museum*, vol. 5, pp. 287–8;

Anon. *The Eccentric Magazine*, vol. 1, pp. 95–6; Anon., *Eccentric Biography; Or, Sketches*, p. 34. On ornamental hermits, see Timbs, *English Eccentrics and Eccentricities*, vol. 1, pp. 156–66.

91. Anon., 'Old Bailey', *The Times* (18 July 1787), p. 3.

92. Anon., 'Prerogative Court', *The Times* (13 April 1826), p. 3.

93. Anon., 'Law Report', *The Times* (8 May 1830), p. 4.

94. Anon., 'Police', *The Times* (18 February 1830), p. 3.

95. Anon., 'Commission of Lunacy', *The Times* (8 August 1825), p. 2.

96. On eccentricity in French psychiatry, see Gill, 'Eccentricity and the Cultural Imagination in Nineteenth-Century France', especially ch. 7.

97. L. Davidoff and C. Hall, *Family Fortunes: Men and Women of the English Middle Class, 1780–1850* (London: Routledge, 1987).

98. A. Vickery, 'Golden Age to Separate Spheres? A Review of the Categories and Chronology of English Women's History', *Historical Journal*, 36 (1993), pp. 383–414.

99. L. E. Klein, 'Gender and the Public/Private Distinction in the Eighteenth Century: Some Questions About Evidence and Analytic Procedure', *Eighteenth-Century Studies*, 29 (1995), pp. 97–109.

100. R. Shoemaker, *Gender in English Society, 1650–1850: The Emergence of Separate Spheres?* (London and New York: Longman, 1998). For an overview of the literature, see, e.g., C. Pateman, *The Disorder of Women: Democracy, Feminism and Political Theory* (Cambridge: Polity Press, 1989), ch. 6.

101. On Snell, see Anon., *Eccentric Biography: Or, Memoirs*, pp. 305–17; Kirby, *Kirby's Wonderful and Eccentric Museum*, vol. 2, pp. 430–38; Wilson, *The Eccentric Mirror*, vol. 1, no. 8, pp. 14–25; Timbs, *English Eccentrics and Eccentricities*, vol. 1, pp. 126–32. On Charke, see Anon., *Eccentric Biography: Or, Memoirs*, pp. 83–5; Timbs, *English Eccentrics and Eccentricities*, vol. 2, pp. 135–7. On Frith, see Anon., *Eccentric Biography: Or, Memoirs*, pp. 128–30. On Talbot, see Kirby, *Kirby's Wonderful and Eccentric Museum*, vol. 2, pp. 160–225; Wilson, *The Eccentric Mirror*, vol. 4, no. 38, pp. 1–22. On Davies, see Kirby, *Kirby's Wonderful and Eccentric Museum*, vol. 3, pp. 420–6. On Langevin, see Kirby, *Kirby's Wonderful and Eccentric Museum*, vol. 6, pp. 105–7. On Bown, see Anon., *Eccentric Biography: Or, Memoirs*, pp. 27–8; Timbs, *English Eccentrics and Eccentricities*, vol. 1, pp. 300–2. On Hammerton, see Kirby, *Kirby's Wonderful and Eccentric Museum*, vol. 4, pp. 311–13. On East, see Kirby, *Kirby's Wonderful and Eccentric Museum*, vol. 3, pp. 414–18; Wilson, *The Eccentric Mirror*, vol. 1, no. 4, pp. 15–22.

102. On Baker, see Anon., *Eccentric Biography: Or, Memoirs*, p. 11; on Fulvia, Anon., *Eccentric Biography: Or, Memoirs*, p. 131; on Queen Christina, Anon., *Eccentric Biography: Or, Memoirs*, pp. 85–9.

103. M. Edgeworth, *Belinda*, 3 vols (London: Printed for J. Thomson, 1801), vol. 2, pp. 170–1.

104. C. Brontë, *Shirley. A Tale,* 3 vols (London: Smith, Elder & Co, 1849), vol. 2, p. 226.

105. C. Dickens, *The Personal History of David Copperfield* (London: Chapman & Hall, 1850), p. 426.

106. C. Yonge, *The Daisy Chain; Or, Aspirations. A Family Chronicle. By the Author of the Heir of Redclyffe, Etc.* (London: John W. Parker & Son, 1856), pp. 175, 176.

107. Anon., *Eccentric Biography; Or, Sketches*, pp. 282–5. For a discussion of transvestism in eighteenth-century England, see M. Kahn, *Narrative Transvestism: Rhetoric and Gender in the Eighteenth-Century English Novel* (Ithaca, NY, and London: Cornell University

Press, 1991). Kahn deals primarily with 'literary transvestism', that is, the use by a male author of a first-person female narrator.

108. On D'Eon, see Anon., *Eccentric Biography; Or, Sketches*, pp. 84–93; Anon., *Eccentric Biography: Or, Memoirs*, pp. 118–22; Fairholt, *Remarkable and Eccentric Characters*, pp. 99–129; Kirby, *Kirby's Wonderful and Eccentric Museum*, vol. 4, pp. 1–29; Wilson, *The Eccentric Mirror*, vol. 4, no. 34, pp. 1–14.

109. Kirby, *Kirby's Wonderful and Eccentric Museum*, vol. 4, p. 1.

110. H. Sussman, *Victorian Masculinities: Manhood and Masculine Poetics in Early Victorian Literature and Art* (Cambridge: Cambridge University Press, 1995), p. 4.

111. Anon., 'Untitled', *The Times* (27 January 1823), p. 2.

112. M. Edgeworth, *Patronage*, 4 vols (London: Printed for J. Johnson & Co, 1814), vol. 1, pp. 41–2.

113. On gender in Burney's novels, see B. Zonitch, *Familiar Violence: Gender and Social Upheaval in the Novels of Frances Burney* (London: Associated University Presses, 1997).

114. F. Burney, *Evelina, or, a Young Lady's Entrance into the World*, 3 vols (London: Printed for T. Lowndes, 1778), vol. 3, p. 72.

115. F. Burney, *Camilla*, 5 vols (London: Printed for T. Payne et al., 1796), vol. 3, pp. 388–9.

116. F. Burney, *The Wanderer; Or, Female Difficulties*, 5 vols (London: Longman et al., 1814), vol. 3, pp. 359–60.

117. F. Burney, *Cecilia, or Memoirs of an Heiress*, 5 vols (London: Printed for T. Payne & Son, 1782), vol. 5, p. 47.

118. On Romantic theories of genius, see, e.g., M. H. Abrams, *The Mirror and the Lamp: Romantic Theory and the Critical Tradition* (1953; Oxford: Oxford University Press, 1971). On genius and the sciences in the Romantic period, see S. Schaffer, 'Genius in Romantic Natural Philosophy', in N. Jardine and A. Cunningham (eds), *Romanticism and the Sciences* (Cambridge: Cambridge University Press, 1990), pp. 82–98. On male genius as effeminacy, see A. Elfenbein, *Romantic Genius: The Prehistory of a Homosexual Role* (New York: Columbia University Press, 1999).

119. T. Carlyle, 'Jean Paul F. Richter', *Edinburgh Review*, 46 (1827), pp. 176–95, at p. 191.

120. W.N., 'Character of Robert Lecky Esq', p. 1.

121. Edgeworth, *Belinda*, vol. 1, pp. 18–19.

122. E. Bulwer Lytton, *Pelham; Or, the Adventures of a Gentleman*, 3 vols (London, 1828), vol. 2, p. 80.

123. On eccentricity and fashion, see Gill, 'Eccentricity and the Cultural Imagination in Nineteenth-Century France', ch. 2.

124. G. Meredith, *The Ordeal of Richard Feverel: A History of Father and Son* (London: Chapman & Hall, 1859), vol. 2, p. 289.

125. Bulwer Lytton, *Pelham*, vol. 2, pp. 64, 197, 89.

126. James Gregory similarly concludes that eccentricity was not solely a privilege of the leisured classes in Gregory, 'Local Characters', p. 165.

127. Gill, 'Eccentricity and the Cultural Imagination in Nineteenth-Century France', p. 24.

128. Ibid., p. 76

129. Quoted in ibid., p. 25.

130. Ibid., p. 344.

131. Hamilton, *Biography*, ch. 4; H. Lee, *Body Parts: Essays in Life-Writing* (London: Chatto & Windus, 2005), pp. 3–4.

132. Brontë, *Shirley*, vol. 1, p. 299. On eccentric femininity, see D. E. Nord, '"Marks of Race"': Gypsy Figures and Eccentric Femininity in Nineteenth-Century Women's Writing', *Victorian Studies*, 41 (1988), pp. 191–210.

133. A notable exception is the self-confessedly eccentric preacher James Kendall: see his *Eccentricity; Or a Check to Censoriousness; with Chapters on Other Subjects* (London: Simpkin, Marshall & Co, 1859).

134. In thinking about self-presentation I have been influenced by E. Goffman, *The Presentation of Self in Everyday Life* (1959; London: Penguin, 1990).

2 Performers, Audiences and Eccentric Identities

1. W. Martin, *A Challenge to the Whole World to Produce a Travelling Machine Equal to this Invention*, broadside dated 26 August 1828 (Newcastle: Edward Walker, 1828), in *Billy Martin Philosopher's Nonsense*, Cambridge University Library, call no. 8000.c.177.57.

2. Anon., *The Martin and the Eagle*, Offprint of article from the *Durham County Advertiser* (11 October 1828), in *Billy Martin Philosopher's Nonsense*, Cambridge University Library, call no. 8000.c.177.33; Anon., 'From the *Tyne Mercury*', Offprint of an article in the *Tyne Mercury*, 6 October (Newcastle upon Tyne: William Mitchell, 1828), in *Billy Martin Philosopher's Nonsense*, Cambridge University Library, call no. 8000.c.177.37.

3. Anon., *The Martin and the Eagle*.

4. Anon., *The Martin and the Eagle*.

5. Anon., *The Martin and the Eagle*.

6. J.H., 'William Martin', *The Northumbrian* (7 April 1883), p. 280.

7. J. B. Langhorne, 'W. Martin, the Natural Philosopher', *Notes and Queries*, 4th Series, 12 (1873), p. 134.

8. Anon., 'William Martin, the Anti-Newtonian Philosopher', *Monthly Chronicle of North-Country Lore and Legend*, 1 (1887), pp. 343–9, at p. 343.

9. J. Gregory, '"Local Characters": Eccentricity and the North-East in the Nineteenth Century', pp. 163–86. On local characters, see also J. Gregory, '"Local Characters" and Local, Regional and National Identities in Nineteenth-Century England and Scotland', pp. 45–60.

10. A Disciple, *A New Song, Called the 'Northumberland Eagle Mail'* (c. 1828), in *Billy Martin Philosopher's Nonsense*, Cambridge University Library, call no. 8000.c.177.32.

11. Biographical sources on Martin include T. Balston, *The Life of Jonathan Martin, Incendiary of York Minster, with Some Account of William and Richard Martin* (London: Macmillan & Co. Ltd, 1945); W. Martin, *A Short Outline of the Philosopher's Life, from Being a Child in Frocks to This Present Day, after the Defeat of All Impostors, False Philosophers, since the Creation; by the Will of the Mighty God of the Universe, He Has Laid the Grand Foundation for Church Reform by True Philosophy &c.* (Newcastle: J. Blackwell and Co., 1833).

12. Cited in W. Feaver, *The Art of John Martin* (Oxford: Clarendon Press, 1975), p. 2.

13. H. Dircks, *Perpetuum Mobile; Or, Search for Self-Motive Power, during the 17th, 18th, and 19th Centuries* (London: E. & F. N. Spon, 1861), pp. i–ii. See also A. De Morgan, *A Budget of Paradoxes* (London: Longmans, Green, & Co, 1872), pp. 285, 342.

14. W. Martin, *A New System of Natural Philosophy, on the Principle of Perpetual Motion*, pp. 15–16.

15. Ibid., pp. 64–7.

16. W. Martin, *The Thunder Storm of Dreadful Forked Lightning: God's Judgment Against All False Teachers, that Cause the People to Err, and Those that Are Led by Them Are Destroyed, According to God's Word. Including an Account of the Railway Phenomenon, the Wonder of the World!* (Newcastle upon Tyne: Pattison & Ross, 1837), in *Billy Martin Philosopher's Nonsense*, Cambridge University Library, call no. 8000.c.177.94, pp. 23–5.

17. Martin, *A Short Outline of the Philosopher's life*, p. 33.

18. R. Welford, *Men of Mark 'Twixt Tyne and Tweed* (London: Walter Scott, 1895), p. 175.

19. G. W. Couper, *Original Poetry* (North Shields: R. Henderson, 1828), p. 7.

20. D. Orange, 'Rational Dissent and Provincial Science: William Turner and the Newcastle Literary and Philosophical Society', in I. Inkster and J. Morrell (eds), *Metropolis and Province: Science in British Culture, 1780–1850* (London: Hutchinson, 1983), pp. 205–30. For an insight into cultural life in Newcastle in the 1820s, see T. G. Wright, *The Diary of Thomas Giordani Wright, Newcastle doctor, 1826–1829* (Woodbridge: Boydell, 2001); J. Uglow, *Nature's Engraver: A Life of Thomas Bewick* (London: Faber, 2006).

21. On science teaching at Mechanics' Institutes, see, e.g., I., Inkster, 'Science and the Mechanics' Institutes, 1820–1850: The Case of Sheffield', *Annals of Science*, 32 (1975), pp. 451–74; S. Shapin and B. Barnes, 'Science, Nature and Control: Interpreting Mechanics' Institutes', *Social Studies of Science*, 7 (1977), pp. 31–74.

22. See local newspaper advertisements, e.g., Anon., 'Queens Head Inn, Morpeth. A Meeting of the Committee Assembled for the Adoption of Measures for Building a New Bridge at Morpeth', *Newcastle Courant* (20 September 1828), p. 1; Anon., 'To Be Sold by Auction at Mr Fletcher's, the Turk's Head Inn, Bigg Market, Newcastle ... A Newly-Built Dwelling House', *Newcastle Courant* (8 September 1827), p. 1; Anon, 'Mechanical and Optical Exhibition, at Mr Fletcher's Long Room, Turk's Head, Bigg Market', *Tyne Mercury* (1 April 1828), p. 3; Anon, 'Now Exhibiting at the Turks Head Long Room, Sinclair & Co's Grand Dioramic or Peristrephic New Panorama of the Battle of Navarino', *Newcastle Courant* (28 June 1828), p. 1. On public houses as sites for the production of scientific knowledge, see Secord, 'Science in the Pub, pp. 269–315.

23. W. B. Scott, *Autobiographical Notes of the Life of William Bell Scott*, ed. W. Minto, 2 vols (London: James R. Osgood, McIlbaine & Co, 1892), vol. 1, p. 197.

24. On John Martin's life, see M. L. Pendered, *John Martin, Painter: His Life and Times* (London: Hurst & Blackett, 1923).

25. *The Paradise Lost of Milton, with Illustrations, Designed and Engraved by John Martin* (London: Septimus Prowett, 1827); G. Mantell, *Wonders of Geology* (London, 1838). On Martin's genre, see M. Paley, *The Apocalyptic Sublime* (New Haven, CT: Yale University Press, 1986).

26. S. C. Hall, *Retrospect of a Long Life: From 1815 to 1883*, 2 vols (London: Richard Bentley & Son, 1883), vol. 2, p. 225. On John Martin's friends, see Pendered, *John Martin*, passim. Much of Pendered's information is gleaned from the reminiscences of John Martin's son, Leopold. While it is possible that Leopold Martin misremembered or overstated the connections of his father, it is nevertheless clear that he moved in high circles and had some influential friends.

27. M. Adams, 'John Martin and the Prometheans', *History Today*, 56:8 (2006), pp. 40–7, at p. 42.

28. Cited in Pendered, *John Martin*, pp. 232–3.

29. On Jonathan's life, see Balston, *Life of Jonathan Martin*; Jonathan Martin, *The Life of Jonathan Martin, of Darlington, Tanner: Written by Himself, etc.* (Darlington: Thomas Thompson, 1825).

30. Cited in Balston, *Life of Jonathan Martin*, p. 48.

31. Ibid., p. 83.

32. Cited in ibid., p. 82.

33. Cited in ibid., p. 100.

34. Martin, *A Short Outline of the Philosopher's Life, from Being a Child in Frocks to This Present Day*, pp. 37–41; W. Martin, *The Christian Philosopher's Explanation of the General Deluge, and the Proper Cause of All the Different Strata: Wherein It Is Clearly Demonstrated, That One Deluge Was the Cause of the Whole, Which Divinely Proves That God Is Not a Liar, but That the Bible Is Strictly True* (Newcastle: Printed by W. Fordyce, 1834), p. 4.

35. Martin, *The Christian Philosopher's Explanation of the General Deluge*, p. 3.

36. Cited in Feaver, *The Art of John Martin*, p. 2.

37. W. Martin, *The Total Defeat of the Yearly Meeting of the British Association of Scientific Asses* (Newcastle: Pattison & Ross, 1838), p. 15.

38. W. Martin, *Martin, Natural Philosopher!* (Newcastle: Edward Walker, 1827), broadside in *Billy Martin Philosopher's Nonsense*, Cambridge University Library, call no. 8000. c.177.40.

39. Fair Play, 'Letter to the Editor of the Tyne Mercury', *Tyne Mercury* (10 April 1827), p. 3.

40. Martin, *New System*, pp. 11–12.

41. Martin, *New System*, pp. 12–13.

42. Martin, *New System*, p. 23. An exception is the spring tides, which according to the Martinian system are caused by the oscillation of the Moon: ibid., p. 12.

43. Martin, *New System*, pp. 22–5. The missing 1 in the sum is due to the way Martin rounds his figures.

44. W. Martin, *The Downfall of the Newtonian False System of Philosophy* (1827), in *Billy Martin Philosopher's Nonsense*, Cambridge University Library, call no. 8000.c.177.11.

45. On advertising in this period see, e.g., N. McKendrick, 'George Packwood and the Commercialization of Shaving: The Art of Eighteenth-Century Advertising or "The Way to Get Money and Be Happy"', in N. McKendrick, J. Brewer and J. H. Plumb (eds), *The Birth of a Consumer Society: The Commercialization of Eighteenth-Century England* (Bloomington: Indiana University Press, 1982), pp. 145–94.

46. W. Martin, 'New System of Philosophy', *Newcastle Courant* (24 May 1828), p. 1.

47. W. Martin, 'New System of Philosophy', *Newcastle Courant* (29 March 1828), p. 1.

48. W. Martin, 'New System of Philosophy', *Newcastle Courant* (14 June 1828), p. 1; William Martin, 'New System of Philosophy', *Newcastle Courant* (21 June 1828), p. 1.

49. W. Martin, 'New System of Philosophy', *Newcastle Courant* (5 April 1828), p. 1.

50. W. Martin, 'New System of Philosophy', *Newcastle Courant* (10 May 1828), p. 1.

51. W. Martin, 'New System of Philosophy', *Newcastle Courant* (31 May 1828), p. 1.

52. W. Martin, 'New System of Philosophy', *Newcastle Courant* (7 June 1828), p. 1.

53. A Disciple, *Martin's Philosophy!* (Newcastle: Edward Walker, 1828), in *Billy Martin Philosopher's Nonsense*, Cambridge University Library, call no. 8000.c.179.54.

54. F.C., *Martinian Philosophy* (1828), in *Billy Martin Philosopher's Nonsense*, Cambridge University Library, call no. 8000.c.179.54.

55. Ibid.

56. W. Martin, 'New System of Philosophy', *Newcastle Courant* (19 January 1828), p. 1.

57. W. Martin, 'New System of Philosophy', *Newcastle Courant* (29 March 1828), p. 1.

58. W. Martin, 'New System of Philosophy', *Newcastle Courant* (19 April 1828), p. 1.

59. Ibid.

60. W. Martin, 'New System of Philosophy', *Newcastle Courant* (5 April 1828), p. 1.

61. W. Martin, 'New System of Philosophy', *Newcastle Courant* (17 May 1828), p. 1.

62. W. Martin, 'New System of Philosophy', *Newcastle Courant* (15 March 1828), p. 1.

63. 'Ecce Homo' identifies the referent of Martin's initial attack as 'Mr H. A.' A later attack on Martin states that Atkinson was one of Martin's key opponents: R. May, *News from Wallsend* (Newcastle, 1827), in *Billy Martin Philosopher's Nonsense*, Cambridge University Library, call no. 8000.c.177.48.

64. Newtonian, *Ecce Homo* (Newcastle: Blagburn, 1827), in *Billy Martin Philosopher's Nonsense*, Cambridge University Library, call no. 8000.c.177.38.

65. W. Martin, *Lines on the Safety Lamp; With a Reply to an Attack Headed 'Ecce Homo', Written by a Fellow Who Calls Himself 'A Newtonian'* (1827), in *Billy Martin Philosopher's Nonsense*, Cambridge University Library, call no. 8000.c.177.45.

66. Anon., *True Philosophy!* (1827), in *Billy Martin Philosopher's Nonsense*, Cambridge University Library, call no. 8000.c.177.51.

67. J. Keats, *Lamia* (1819; London: Kessinger Publishing, 2004), pp. 18–19.

68. See, e.g., A. Cunningham and N. Jardine (eds), *Romanticism and the Sciences* (Cambridge: Cambridge University Press, 1990).

69. Cited in J. Sysak, 'Coleridge's Construction of Newton', *Annals of Science*, 50 (1993), pp. 59–81, at p. 60.

70. G. Finley, 'The Deluge Pictures: Reflections on Goethe, J. M. W. Turner and Early Nineteenth-Century Science', *Zeitschrift für Kunstgeschichte*, 60 (1997), pp. 530–48, at p. 530.

71. Cited in Finley, 'Deluge Pictures', p. 534.

72. J. Cage, *Colour and Culture: Practice and Meaning from Antiquity to Abstraction* (Thames & Hudson, 1993), p. 203.

73. Martin, *New System of Philosophy*, pp. 34–5.

74. F. Reid, 'Isaac Frost's *Two Systems of Astronomy* (1846): Plebeian Resistance and Scriptural Astronomy', *British Journal for the History of Science*, 38 (2005), pp. 161–77.

75. Ibid., p. 169.

76. See N. Aston, 'From Personality to Party: The Creation and Transmission of Hutchinsonianism, *c.* 1725–1750', *Studies in History and Philosophy of Science*, 35 (2004), pp. 625–44.

77. Ibid., p. 627.

78. Cited in ibid., p. 628.

79. Ibid, p. 626.

80. On Hutchinsonianism in nineteenth-century Hackney, London, see P. Corsi, *Science and Religion: Baden Powell and the Anglican Debate, 1800–1860* (Cambridge: Cambridge University Press, 1988), p. 23.

81. Martin, *A Short Outline of the Philosopher's Life*, p. 47.

82. W. Martin, *The Defeat of Learned Humbugs, and the Downfall of All False Philosophers, in the Nineteenth Century, for the Good of All Mankind, and the Christian Church* (Newcastle upon Tyne, 1832), p. 49.

83. Martin, *The Total Defeat of the Yearly Meeting of the British Association of Scientific Asses*, p. 15.

84. Ibid., p. 15.

85. C. Webster, *The Great Instauration: Science, Medicine and Reform 1626–1660* (London: Duckworth, 1975), p. 23.

86. Ibid., pp. 21–7.

87. Martin, *The Christian Philosopher's Explanation of the General Deluge, and the Proper Cause of All the Different Strata*, pp. 10–11.
88. Ibid., p. 11.
89. W. Martin, *Prophetic Knowledge from That Spirit of God which Directed Daniel the Prophet to Interpret the Hand Writing on the Wall to the Impious King Belshazzar* (Newcastle: Pattison & Ross, 1839), p. 1.
90. Martin, *The Defeat of Learned Humbugs, and the Downfall of All False Philosophers*, p. 48.
91. Harrison, *The Second Coming*, p. 6. Compare M. Paley, who distinguishes between varieties of millennial belief, in *Apocalypse and Millennium in English Romantic Poetry* (Oxford: Clarendon Press, 1999), p. 3. On millenarianism, see also I. McCalman, *Radical Underworld: Prophets, Revolutionaries and Pornographers in London, 1795–1840* (Cambridge: Cambridge University Press, 1988).
92. I. McCalman, 'New Jerusalems: Prophecy, Dissent and Radical Culture in England, 1786–1830', in K. Haakonssen (ed.), *Enlightenment and Religion: Rational Dissent in Eighteenth-Century Britain* (Cambridge: Cambridge University Press, 1997), pp. 312–35, at p. 317; K. Knox, 'Lunatick Visions: Prophecy, Signs and Scientific Knowledge in 1790s London', *History of Science*, 37 (1999), pp. 427–58, at p. 430.
93. R. Brothers, *A Revealed Knowledge of the Prophecies and Times. Book the First. Wrote under the Direction of the Lord God, and Published by His Sacred Command; It Being the First Sign of Warning, for the Benefit of All Nations. Containing, with Other Remarkable Things, Not Revealed to Any Other Person on Earth, the Restoration of the Hebrews to Jerusalem, by the Year of 1798, under Their Revealed Prince and Prophet, Richard Brothers*, 2 vols (London, 1794).
94. Harrison, *The Second Coming*, p. 61.
95. Cited in ibid., p. 63.
96. Knox, 'Lunatick Visions', p. 436.
97. Harrison, cited in Knox, 'Lunatick visions', p. 435.
98. Knox, 'Lunatick visions', p. 435.
99. Martin, *The Downfall of the Newtonian System*.
100. Cited in Harrison, *The Second Coming*, p. 65.
101. W. Martin, *William Martin's Challenge to the Whole Terrestrial Globe, as a Philosopher and Critic, and Poet and Prophet; Shewing the Travels of His Mind, the Quick Motion of the Soul, that Never-Dying Principle, the Spirit Belonging to Mortal Man* (Newcastle upon Tyne: Printed for the author by Wm Fordyce, 1829), p. 1.
102. Ibid., p. 1.
103. W. Martin, *Belshazzar's Feast*, broadside dated 26 December 1826 (Newcastle: Edward Walker), in *Billy Martin Philosopher's Nonsense*, Cambridge University Library, call no. 8000.c.177.43.
104. Harrison, *The Second Coming*, p. 47.
105. Knox, 'Lunatick Visions', p. 427.
106. Cited in Harrison, *The Second Coming*, p. 51. On nineteenth-century astrology, see also Patrick Curry, *A Confusion of Prophets*.
107. On almanacs, see, e.g., K. Anderson, 'The Weather Prophets: Science and Reputation in Victorian Meteorology', *History of Science*, 37 (1999), pp. 178–216; M. Perkins, *Visions of the Future*.
108. Harrison, *The Second coming*, p. 44.
109. Knox, 'Lunatick Visions', p. 428.

110. Ibid., p. 430.

111. Ibid. pp. 441–2.

112. M. Bakhtin, *Rabelais and His World*, trans. Helen Iswolsky (first published in Russian 1965; Cambridge, MA: Massachuetts Institute of Technology, 1968), p. 4.

113. Ibid., p. 6.

114. Ibid., p. 4.

115. Ibid., p. 10.

116. On English customs, see, e.g., B. Bushaway, *By Rite: Custom, Ceremony and Community in England 1700–1880* (London: Junction Books, 1982).

117. See, e.g., M. Ingram, 'Ridings, Rough Music and Mocking Rhymes in Early Modern England', in B. Reay (ed.), *Popular Culture in Seventeenth-Century England* (London: Croom Helm, 1985), pp. 166–97; E. P. Thompson, *Customs in Common* (London: The Merlin Press, 1991), ch. 8.

118. T. Hardy, *Mayor of Casterbridge*, ch. 39, cited in Thompson, *Customs in Common*, p. 469.

119. Anon., *The Martin and the Eagle*.

120. Couper, *Original Poetry*, p. 20.

121. Ibid., p. 20.

122. Ingram, 'Ridings, Rough Music and Mocking Rhymes', p. 178.

123. Anon., *Imperial Mandate* (Newcastle: Blagburn, 1827), in *Billy Martin Philosopher's Nonsense*, Cambridge University Library, call no. 8000.c.177.46. Martin attributes 'Imperial Mandate' to Atkinson in W. Martin, *A New Philosophical Song or Poem Book, called the Northumberland Bard; Or, the Downfall of All False Philosophy* (Newcastle: Thomas Blagburn, 1827), p. 31.

124. Thompson, *Customs in Common*.

125. Knox, 'Lunatick Visions', p. 430.

126. May, *News from Wallsend*. MacKenzie was also a local historian: see his *Descriptive and Historical Account of the Town and County of Newcastle Upon Tyne: Including the Borough of Gateshead* (Newcastle upon Tyne: Mackenzie and Dent, 1827).

127. Orange, 'Rational Dissent and Provincial Science', p. 206.

128. Ibid., p. 221.

129. B. Hilton, *A Mad, Bad and Dangerous People?: England 1783–1846* (Oxford: Clarendon Press, 2006), p. 463.

130. Orange, 'Rational Dissent and Provincial Science', p. 217.

131. Ibid., p. 221.

132. C. J. Hunt, *The Book Trade in Northumberland and Durham to 1860* (Newcastle upon Tyne: Thorne's Students' Bookshop, 1975). On the print trades in Newcastle, see also C. J, Hunt and P. C. G. Isaac, 'The Regulation of the Booktrade in Newcastle Upon Tyne at the Beginning of the Nineteenth Century', *Archaeologia aeliana*, 5:5 (1977), pp. 163–78; P. J. Wallis, *The Book Trade in Northumberland and Durham to 1860: A Supplement to C. J. Hunt's Biographical Dictionary* (Newcastle upon Tyne: Thorne's Bookshops, 1981).

133. W. Martin, *A New Philosophical Song or Poem Book, called the Northumberland Bard*.

134. See, e.g., Anon., *Triumph of Truth* (North Shields: Pollock, 1828), in *Billy Martin Philosopher's Nonsense*, Cambridge University Library, call no. 8000.c.177.52.

135. Anon., *Astronomy* (1828), in *Billy Martin Philosopher's Nonsense*, Cambridge University Library, call no. 8000.c.177.35; Anon, Untitled Advertisement for Mr Lloyd's lecture, *Tyne Mercury* (27 May 1828), p. 3; Anon., *Mr Walker's Lecture on Astronomy*, *Tyne Mercury* (27 May 1828), p. 3.

136. Anon., *Martin Victorious!* (Newcastle: Edward Walker, 1828), in *Billy Martin Philosopher's Nonsense*, Cambridge University Library, call no. 8000.c.177.55.
137. W. Martin, 'New System of Philosophy', *Newcastle Courant* (24 May 1828), p. 1.
138. Ibid., p. 1.
139. Gregory, '"Local Characters": Eccentricity and the North-East in the Nineteenth Century', pp. 176–7.
140. F. Gossman, *Past Events: An Every-Day Register of Events of Local and General Public Interest, 1880* (Newcastle, 1881), p. 168, cited in Gregory, '"Local Characters": Eccentricity and the North-East in the Nineteenth Century', p. 177.
141. Fair Play, 'Letter to the Editor of the Tyne Mercury', *Tyne Mercury* (3 April 1827), p. 4.
142. *Tyne Mercury* (19 May 1812), cited in M. Milne, 'Periodical Publishing in the Provinces: The Mitchell Family of Newcastle upon Tyne', *Victorian Periodicals Newsletter*, 10 (1977), pp. 174–82, at p. 176.
143. Fair Play, 'Letter' (3 April 1827).
144. Ibid.
145. Ibid.
146. Fair Play, 'Letter' (10 April 1827), p. 3.
147. M. Chase, *'The People's Farm': English Radical Agrarianism 1775–1840* (Oxford: Clarendon Press, 1988), ch. 4. See also H. T. Dickinson, 'Spence, Thomas (1750–1814)', *Oxford Dictionary of National Biography*.
148. Chase, *'The People's Farm'*, p. 56.
149. A Real Disciple, 'On Reading the Works of Mr Martin, N. P. & P.', in Martin, *A New Philosophical Song or Poem Book, called the Northumberland Bard*, p. 30.
150. For 'Great Martin', see Anon., *True Philosophy!*; for 'Professor Martin', Martin, *Martin, Natural Philosopher!*; for 'Champion of Truth' and 'god-like man', Martin, 'New System of Philosophy', *Newcastle Courant* (24 May 1828), p. 1; for 'genius is divine', Martin, *A New Philosophical Song or Poem Book, Called the Northumberland Bard*, p. 30.
151. See, e.g., Couper, *Original Poetry*, p. 6.
152. Martin, *New Philosophical Song Book or Poem Book, Called the Northumberland Bard*, p. 30; A Disciple, *A New Song, Called the 'Northumberland Eagle Mail'*.
153. Martin, *Martin, Natural Philosopher!*; F.C., *Martinian Philosophy*.
154. M. Gluckman, 'Rituals of Rebellion in South-East Africa', in Gluckman (ed.), *Order and Rebellion in Tribal Africa* (London: Cohen & West, 1963), pp. 110–36, at pp. 125–6.
155. V. Turner, *The Anthropology of Performance* (1987; New York: PAJ Publications, 1988), pp. 24–5.
156. V. Turner, *The Ritual Process: Structure and Anti-Structure* (London: Routledge & Keegan Paul, 1969), p. 176. On rituals of inversion in a historical context, see, e.g., C. Hill, *The World Turned Upside Down: Radical Ideas During the English Revolution* (1972; London: Penguin Books, 1991); N. Z. Davies, *Society and Culture in Early Modern France: Eight Essays* (1965; Stanford, CA: Stanford University Press, 1975).
157. Davies, *Society and Culture in Early Modern France*.
158. Stallybrass and White, *The Politics and Poetics of Transgression*, p. 14.
159. Hall, *Retrospect of a Long Life*, vol. 2, p. 227.
160. Anon., 'William Martin, the Natural Philosopher!', *The Newcastle Magazine*, 1 (1822), pp. 25–9. For Martin's response to the review, see Martin, 'New System of Philosophy', *Newcastle Courant* (24 May 1828), p. 1.
161. Ibid., pp. 27–8.
162. Ibid., p. 27.
163. Ibid., p. 27.

3 'Beyond the Pale of Ordinary Criticism'

1. Hawkins, *The Book of the Great Sea-Dragons.* Hereafter referred to as *Sea-Dragons.* On Hawkins's eccentricity, see also, V. Carroll, '"Beyond the Pale of Ordinary Criticism": Eccentricity and the Fossil Books of Thomas Hawkins', *Isis*, 98 (2007), pp. 225–65.

2. Anon., Review of *Sea-Dragons, New Monthly Magazine*, p. 429.

3. T. Hawkins, *Memoirs of Ichthyosauri and Plesiosauri, Extinct Monsters of the Ancient Earth, with Twenty-Eight Plates Copied from Specimens in the Author's Collection of Fossil Organic Remains* (London: Relfe & Fletcher, 1834). Hereafter referred to as *Memoirs.* A second edition of *Memoirs*, with some additions, was produced in 1835 (although it was dated 1834 on the title page). All further references are to the first edition unless otherwise stated.

4. Anon., Review of *Memoirs, The Literary Gazette and Journal of Belles Lettres*, 910 (1834), pp. 444–5, at p. 444; Anon., Review of *Memoirs, New Monthly Magazine*, 42 (1834), pp. 240–2, at p. 240. The second review was probably written by the editor of the *New Monthly Magazine*, S. C. Hall.

5. Anon., Review of *Memoirs, New Monthly Magazine*, p. 240; Anon., Review of *Memoirs, The Literary Gazette*, p. 445.

6. Anon., Review of *Sea-Dragons, New Monthly Magazine*, p. 431.

7. An exception is R. O'Connor, 'Thomas Hawkins and Geological Spectacle', *Proceedings of the Geologists' Association*, 114 (2003), pp. 227–41. See also O'Connor, 'The Poetics of Geology: A Science and its Literature in Britain, 1802–1856' (PhD dissertation: University of Cambridge, 2003), passim; O'Connor, *The Earth on Show: Fossils and the Poetics of Popular Science, 1802–1856* (Chicago, IL: Chicago University Press, 2008). O'Connor does not analyse Hawkins's books specifically in terms of their perceived eccentricity. However I have found his work extremely helpful, especially as regards placing Hawkins's books within the wider context of geological literature in the period.

8. R. W. Purcell and S. J. Gould, *Finders, Keepers: Eight Collectors* (London: Hutchinson Radius, 1992), pp. 95, 107.

9. Hawkins's tendency to mix genres has been noted by O'Connor, 'Poetics of Geology', ch. 4. However, O'Connor does not analyse this explicitly in connection with the perceived eccentricity of Hawkins's works.

10. Although many variations and some alternatives have been offered, this basic formulation is widely accepted. Works on genre that I have found helpful include Genette, *Paratexts*; D. McKenzie, *Bibliography and the Sociology of Texts* (London: The British Library, 1986); M. Riffaterre, *Text Production*, trans. T. Lyons (first published in French 1979; New York: Columbia University Press, 1983); T. Todorov, *The Fantastic: A Structural Approach to a Literary Genre* (Ithaca, NY: Cornell University Press, 1975). For an alternative, ahistorical theory of genre, see, e.g., N. Frye, *Anatomy of Criticism: Four Essays* (Princeton, NJ: Princeton University Press, 1957), pp. 131–223. Frye argues that genres are 'archetypal' manifestations of universal, unchanging characteristics of the human imagination.

11. See especially Genette, *Paratexts*. On the rhetorical analysis of scientific writing, see e.g. C. Bazerman, *Shaping Written Knowledge: The Genre and Activity of the Experimental Article in Science* (Madison, WI: The University of Wisconsin Press, 1988); A. Benjamin et al. (eds), *The Figural and the Literal: Problems of Language in the History of Science and Philosophy, 1630–1800* (Manchester: Manchester University Press, 1987); P. Dear (ed.), *The Literary Structure of Scientific Argument* (Philadelphia, PA: University of Pennsylvania Press, 1991); A. Gross, *The Rhetoric of Science* (Cambridge, MA: Harvard University

Press, 1990); G. Myers, *Writing Biology: Texts in the Social Construction of Scientific Knowledge* (Madison, WI: University of Wisconsin Press, 1990).

12. Hawkins, *Sea-Dragons*, pp. 15–16.

13. Hawkins, *Memoirs*, p. iii.

14. [Percival Lord], Review of *Memoirs*, *Athenaeum*, 347 (1834), pp. 469–70, at p. 469.

15. Hawkins's medical training is discussed in M. Taylor, 'Hawkins, Thomas (1810–1899)', *Oxford Dictionary of National Biography* (Oxford: Oxford University Press, 2004). Hawkins mentions his father's death in *Memoirs*, p. v. On Hawkins's life and collecting practices, see A. Bulleid, 'Notes on the Life and Work of Thomas Hawkins F.G.S.', *Somerset Archaeology and Natural History*, 89 (1943), pp. 59–71; C. McGowan, *The Dragon Seekers* (Cambridge, MA: Perseus Publishing, 2001); M. Taylor, 'Thomas Hawkins FGS', *Geological Curator*, 5 (1987), pp. 112–14; M. Taylor, 'Thomas Hawkins of "The Great Sea Dragons": The Isle of Wight Connection', *Geological Society of the Isle of Wight Newsletter*, 1:5 (1997), pp. 5–10. See also Hawkins's fragmentary autobiography, written late in life, *My Life and Works Etc.* (London: The Chiswick Press, 1887).

16. Hawkins, *Memoirs*, 2nd edn, p. 53.

17. Gideon Mantell to Benjamin Silliman, 18 June 1834. Cited in D. Dean, *Gideon Mantell and the Discovery of Dinosaurs* (Cambridge: Cambridge University Press, 1999), p. 140.

18. H. Torrens, 'Mary Anning (1799–1847) of Lyme: "The Greatest Fossilist the World Ever Knew"', *British Journal for the History of Science*, 28 (1995), pp. 257–84, at p. 259. Also see S. Howe, T. Sharpe and H. Torrens, *Ichthyosaurs: A History of Fossil 'Sea-Dragons'* (Cardiff: National Museum of Wales, 1981).

19. G. Cuvier, *Recherches sur les ossemens fossiles, où l'on rétablit les caractères de plusieurs animaux dont les révolutions du globe ont détruit les espèces*, 5 vols (Paris: G. Dufour & E. D'Ocagne, 1821–4), vol. 5, part 2, p. 444.

20. E. Home, 'Some Account of the Fossil Remains of an Animal More Nearly Allied to Fishes than Any of the Other Classes of Animals', *Philosophical Transactions of the Royal Society* (1814), pp. 571–7. In 1819, Home changed his mind and named the animal 'Proteosaurus', arguing that it was a link between reptiles and the recently-discovered *Proteus*, but his argument was rejected. See E. Home, 'An Account of the Fossil Skeleton of the Proteo-Saurus', *Philosophical Transactions of the Royal Society* (1819), pp. 209–11; E. Home, 'Reasons for Giving the Name Proteo-Saurus to the Fossil Skeleton Which Has Been Described', *Philosophical Transactions of the Royal Society* (1819), pp. 212–16.

21. H. De la Beche and W. Conybeare, 'Notice of the Discovery of a New Fossil Animal, Forming a Link Between the Ichthyosaurus and Crocodile, Together With General Remarks on the Osteology of the Ichthyosaurus', *Transactions of the Geological Society*, 5 (1821), pp. 559–94, at p. 562.

22. Ibid., p. 560.

23. W. Buckland, 'Notice on the Megalosaurus or Great Fossil Lizard of Stonesfield', *Transactions of the Geological Society of London*, 2:1 (1824), pp. 390–6. G. Mantell, 'Notice on the Iguanodon, a Newly Discovered Fossil Reptile, from the Sandstone of Tilgate Forest, in Sussex', *Philosophical Transactions of the Royal Society of London*, 115 (1825), pp. 179–86.

24. Hawkins, *Memoirs*, 2nd edn, p. 53.

25. Hawkins, *Memoirs*, p. 32.

26. Ibid., p. 25.

27. Ibid., pp. 12–13.

28. Hawkins, *Memoirs*, 2nd edn, p. 53.

29. Ibid., p. 53.

30. McGowan, *The Dragon Seekers*, pp. 127–37.
31. Anning quote, cited in McGowan, *The Dragon seekers*, p. 139; E. Curwen (ed.), *The Journal of Gideon Mantell Surgeon and Geologist Covering the Years 1818–1852* (London: Oxford University Press, 1940), p. 111.
32. Hawkins writes that his specimen was prepared by 'a trusty Lucchese', *Memoirs*, p. 13. This is presumably the Italian sculptor referred to by Gideon Mantell in a letter to Benjamin Silliman of 18 January 1834, cited in Dean, *Gideon Mantell*, p. 144n.
33. Hawkins, *Memoirs*, 2nd edn, p. 54.
34. Curwen, *The Journal of Gideon Mantell*; R. Owen, *The Life of Richard Owen*, 2 vols (London: John Murray, 1894), vol. 1, pp. 165–6.
35. Dean, *The Journal of Gideon Mantell*, p. 139.
36. Hawkins, *Memoirs*, 2nd edn, p. 54.
37. On the *Gentleman's Magazine*, see C. Darcy, 'The *Gentleman's Magazine*', in A. Sullivan (ed.), *British Literary Magazines: The Romantic Age 1789–1836* (Westport, CT, and London: Greenwood Press, 1983), pp. 136–140.
38. R. Duncan, 'The *Literary Gazette*', in Sullivan, *British Literary Magazines*, pp. 242–6.
39. D. Spurgeon, 'The *Athenaeum*', in Sullivan, *British Literary Magazines*, pp. 21–4.
40. W. Houghton (ed.), *The Wellesley Index to Victorian Periodicals* (Toronto: University of Toronto Press, 1979), vol. 3, pp. 161–72; L. Schachterle, 'The *Metropolitan Magazine*', in Sullivan, *British Literary Magazines*, pp. 304–8.
41. R. Yeo, *Defining Science: William Whewell, Natural Knowledge, and Public Debate in Early Victorian Britain* (Cambridge: Cambridge University Press, 1993), pp. 77–87.
42. Duncan, 'The *Literary Gazette*', p. 243; Spurgeon, 'The *Athenaeum*', p. 21; Houghton, *The Wellesley Index*, vol. 3, p. 166.
43. Hawkins, *Memoirs*, 2nd edn, p. viii.
44. E. Copleston, *Advice to a Young Reviewer, with a Specimen of the Art* (Oxford, 1807), p. 2. This satirical and extremely funny pamphlet was produced anonymously; authorship is attributed in Yeo, *Defining Science*, p. 78, and elsewhere. On contemporary intellectual debates about the nature of criticism, see, e.g., M. H. Abrams, *The Mirror and the Lamp*; J. Klancher, 'British Theory and Criticism: Romantic Period and Early Nineteenth Century', in M. Groden and M. Kreiswirth, (eds), *The Johns Hopkins Guide to Literary Theory and Criticism* (Baltimore, MD: Johns Hopkins University Press, 1997), pp. 112–15. On reviewing, see P. Parrinder, *Authors and Authority: A Study of English Literary Criticism and its Relation to Culture 1750–1900* (London: Routledge & Keegan Paul, 1977); W. Rowland, *Literature and the Marketplace: Romantic Writers and their Audiences in Great Britain and the United States* (Lincoln, NE, and London: University of Nebraska Press, 1996).
45. Genette, *Paratexts*, p. 1.
46. In thinking about the role of preconceptions in shaping interpretations I have been influenced by H.-G. Gadamer, *Truth and Method*, trans. J. Weinsheimer and D. G. Marshall (1960; London: Sheed & Ward, 1989).
47. Hawkins, *Memoirs*, p. v.
48. Genette, *Paratexts*, p. 75.
49. T. Ashe, *Memoirs of Mammoth: And Various Other Extraordinary and Stupendous Bones, of Incognita, or Non-Descript Animals: Found in the Vicinity of the Ohio, Etc.* (Liverpool: Printed by G. F. Harris, 1806); T. Weaver, 'Memoir on the Geological Relations of the East of Ireland', *Transactions of the Geological Society*, 5 (1821), pp. 117–304; P. Martin, *A Geo-*

logical Memoir on Part of Western Sussex; With Some Observations upon Chalk-Basins, the Weald-Denudation, and Outliers-by-Protrusion (London, 1828); Hawkins, *Memoirs*, p. v.

50. Anon., 'Memoirs of Ichthyosauri and Plesiosauri. By T. Hawkins', *Metropolitan Magazine*, 11 (1834), Appendix pp. 1–2, at p. 1.

51. This information was provided by A. Mussell, Archivist to the Geological Society, in personal correspondence.

52. McGowan, *Dragon Seekers*, p. 133.

53. Mantell to Silliman, February 1833; Mantell to Silliman, January 1834. Cited in Dean, *Gideon Mantell*, p. 139.

54. Hawkins, *Memoirs*, p. 37.

55. Hawkins to Buckland, 2 July 1834. Reprinted in Hawkins, *Memoirs*, 2nd edn, p. 55.

56. On Benett, see, e.g., H. Torrens et al., 'Etheldred Benett of Wiltshire, England, the First Lady Geologist', *Proceedings of the Academy of Natural Sciences of Philadelphia*, 150 (2000), pp. 59–123; on the Philpot sisters, see, e.g. J. M. Edmonds, 'The Fossil Collection of the Misses Philpot of Lyme Regis, *Proceedings of the Dorset Natural History and Archaeological Society*, 98 (1976), pp. 43–8.

57. I am grateful to Mike Taylor for helping me to identify some of the subscribers on the list. Thomas Porch Porch is mentioned in M. Taylor, 'Joseph Clark III's Reminiscences about the Somerset Fossil Reptile Collector Thomas Hawkins (1810–1889)', *Somerset Archaeology and Natural History*, 146 (2002), pp. 1–10, at p. 7.

58. I am grateful to Mike Taylor and Jack Morrell for pointing this out to me.

59. O'Connor, *Poetics of Geology*, pp. 126–42.

60. H. De la Beche, *A Geological Manual* (London, 1831); John Phillips, *A Guide to Geology* (London: Longman et al., 1834).

61. Anon., 'Mineral Kingdom', *The Penny Magazine of the Society for the Diffusion of Useful Knowledge*, 2 (1833), pp. 347–9.

62. Gideon Mantell to Benjamin Silliman, 18 June 1834. Cited in Dean, *Gideon Mantell*, pp. 139–40.

63. Anon., *Magazine of Natural History*, 7(1834), pp. 476–9, at p. 477.

64. Cuvier, *Ossemens fossiles*; De la Beche and Conybeare, 'Notice of the Discovery of a New Fossil Animal'.

65. Hawkins, *Memoirs*, p. 43.

66. De la Beche and Conybeare, 'Notice of the Discovery of a New Fossil Animal', p. 582.

67. R. Barthes, *Writing Degree Zero*, trans. A. Lavers and C. Smith (London: Cape, 1967).

68. Cuvier, *Ossemens fossiles*, vol. 5, part 2, p. 458.

69. De la Beche and Conybeare, 'Notice on the Discovery of a New Fossil Animal', p. 582.

70. Anon., *The Literary Gazette* (1834), p. 445.

71. Anon., *Magazine of Natural History* (1834), pp. 477, 478.

72. [W. Weddall], Review of *Memoirs*, *Gentleman's Magazine*, 2 (1834), pp. 121–4, at p. 123. On Weddall, see Anon., 'William Langstaff Weddall', *Gentleman's Magazine* (1851), p. 446.

73. Hawkins, *Memoirs*, 2nd edn, p. 53.

74. Hawkins, *Memoirs*, p. v.

75. E. Gordon, *The Life and Correspondence of William Buckland* (London: John Murray, 1894), p. 26. On the heroic in nineteenth-century geology, see M. Sommer, 'The Romantic Cave? The Scientific and Poetic Quests for Subterranean Spaces in Britain', *Earth Sciences History*, 22 (2003), pp. 172–208.

76. Hawkins, *Memoirs*, p. 33.

77. On the historical development of the pastoral, see J. Sambrook, *English Pastoral Poetry* (Boston, MA: Twayne Publishers, 1983). On the pastoral in nineteenth-century poetry, see J. Barrell, *The Idea of Landscape and the Sense of Place 1730–1840: An Approach to the Poetry of John Clare* (Cambridge: Cambridge University Press, 1972); J. Bate, *Romantic Ecology: Wordsworth and the Environmental Tradition* (London: Routledge, 1991).

78. Hawkins, *Memoirs*, p. 33.

79. Ibid., p. 33.

80. J. Clare, *The Village Minstrel and Other Poems* (London: Taylor & Hessey, 1821).

81. Hawkins, *Memoirs*, p. 33.

82. Ibid., p. 33.

83. [Weddall], Review of *Memoirs*, p. 123.

84. Anon., Review of Hawkins, 'The Book of the Great Sea-Dragons, Ichthisauri, and Pleriosuri [Sic.] Etc.' *Metropolitan Magazine*, 28 (1840), pp. 114–15.

85. Anon., Review of *Memoirs*, *Metropolitan Magazine*, Appendix p. 2.

86. Anon., Review of *Memoirs*, *New Monthly Magazine*, p. 241.

87. Conybeare to Buckland, 4 July 1834, quoted in Howe, Sharpe and Torrens, *Ichthyosaurs*, p. 22.

88. H. Miller, *The Old Red Sandstone; Or, New Walks in an Old Field* (Edinburgh: John Johnstone, 1841). On Miller, see M. A. Taylor, *Hugh Miller: Stonemason, Geologist, Writer* (Edinburgh: National Museums Scotland, 2007).

89. O'Connor, *The Earth on Show*, p. 177.

90. [Weddall], Review of *Memoirs*, p. 123.

91. Anon., 'Hawkins, T., F.G.S.: Memoirs of Ichthyosauri and Plesiosauri Etc.' *Magazine of Natural History*, 7 (1834), pp. 476–9, at p. 477.

92. Hawkins, *Memoirs*, p. 33.

93. M. Rudwick, *Scenes from Deep Time*.

94. Altick, *The Shows of London*, p. 136 and passim.

95. The lithograph was made by the engraver George Scharf, based on a watercolour by De la Beche: see Rudwick, *Scenes from Deep Time*, pp. 42–8. A few years earlier, in 1822, Conybeare produced a lithographed cartoon depicting William Buckland entering an antediluvian hyena's den: see ibid., p. 41. This depicted only one species (in addition to the human time-traveller), and was intended as a joke rather than a serious representation of the prehistoric world.

96. Anon., 'Organic Remains Restored', *Penny Magazine*, 2 (1833), pp. 409–10.

97. Rudwick, *Scenes from Deep Time*, p. 65.

98. Hawkins, *Memoirs*, p. 51.

99. Marianne Sommer has used the metaphor of the conjurer to describe the 'gift' which was sometimes perceived to be necessary for making sense of fossil-containing caves and using them to evoke the deep past. See Sommer, 'The Romantic Cave?', p. 193.

100. On perceived affiliations between natural history and antiquarianism, see, e.g., M. Rudwick, 'The Emergence of a Visual Language for Geological Science, 1760–1840', *History of Science*, 14 (1976), pp. 149–195; J. Paradis, 'The Natural Historian as Antiquary of the World: Hugh Miller and the Rise of Literary Natural History', in M. Shortland (ed.), *Hugh Miller and the Controversies of Victorian Science* (Oxford: Clarendon Press, 1996), pp. 122–50.

101. C. Lyell, *Principles of Geology* (1930–3; London; Penguin, 1997), p. 7.

102. Hawkins, *Sea-Dragons*, p. 15.

103. Ibid., p. 15.

104. Ibid., p. 15.

105. Ibid., p. 15.

106. M. Rudwick, 'The Shape and Meaning of Earth History', in D. C. Lindberg and R. L. Numbers (eds), *God and Nature: Historical Essays on the Encounter between Christianity and Science* (Berkeley: University of California Press, 1986), pp. 296–321, at p. 312.

107. Lyell in a lecture delivered at King's College London, 4 May 1832; cited in J. Moore, 'Geologists and Interpreters of Genesis in the Nineteenth Century', in Lindberg and Numbers (eds), *God and Nature*, pp. 322–50, at p. 337. On the professionalization of geology, see, e.g., R. Porter, 'Gentlemen and Geology', pp. 809–36.

108. J. Sutcliffe, *A Short introduction to the Study of Geology; Comprising a New Theory of the Elevation of the Mountains and the Stratification of the Earth: In Which the Mosaic Account of the Creation and the Deluge is Vindicated* (London, 1817–19); G. Penn, *Comparative Estimate of the Mineral and Mosaical Geologies* (London, 1822); G. Bugg, *Scriptural Geology; Or, Geological Phenomena Consistent Only with the Literal Interpretation of the Sacred Scriptures, Upon the Subjects of the Creation and Deluge* (London: Hatchard & Son, 1826–7); all cited in N. Rupke, *The Great Chain of History: William Buckland and the English School of Geology (1814–1849)* (Oxford: Clarendon Press,1983), pp. 43–7.

109. W. Buckland, *Reliquiae diluvianea: Or, Observations on the Organic Remains Contained in Caves, Fissures, and Diluvial Gravel, and on Other Geological Phenomena, Attesting the Action of an Universal Deluge* (London: John Murray, 1823). Criticism of Buckland is discussed in Rudwick, 'The Shape and Meaning of Earth History', p. 313. On diluvialism, see, e.g., Rupke, *The Great Chain of History*, pp. 29–107. On Buckland's thoughts on the relationship between science and religion, see his *Vindiciae geologicae: Or, the Connexion of Geology with Religion* (Oxford: University Press, 1820), and his Bridgewater Treatise, *Geology and Mineralogy Considered with Reference to Natural Theology*, 2 vols (London: William Pickering, 1836).

110. Rudwick, 'The Shape and Meaning of Earth History', p. 312.

111. J. Brooke and G. Cantor, *Reconstructing Nature* (Edinburgh: T&T Clark, 1998), pp. 57–64; O'Connor, *Poetics of Geology*, ch. 2.

112. I owe this point to Ralph O'Connor.

113. Hawkins, *Sea-Dragons*, pp. 15–16.

114. Anon., Review of *Memoirs*, *Literary Gazette*, p. 445.

115. Booth, *The Rhetoric of Fiction*. See also Foucault, 'What is an Author?', in D. F. Bouchard (ed.), *Language, Counter-Memory, Practice: Selected Essays and Interviews* (Oxford: Basil Blackwell, 1977), pp. 113–38. On the implied author in scientific writing, see M. Coney, 'The Implied Author in Technical Discourse', *Journal of Advanced Composition*, 5 (1988), pp. 163–72.

116. [J. S. Mill], 'The Spirit of the Age', p. 239.

117. Hawkins, *Memoirs*, p. 29.

118. Ibid., p. 29.

119. O'Connor, 'Thomas Hawkins and Geological Spectacle', p. 235.

120. Anon., Review of *Sea-Dragons*, *New Monthly Magazine*, p. 430; Anon., Review of *Sea-Dragons*, *Metropolitan Magazine*, p. 114.

121. Anon., Review of *Memoirs*, *Metropolitan Magazine*, Appendix p. 2.

122. Anon., Review of *Sea-Dragons*, *Metropolitan Magazine*, p. 114.

123. Anon., Review of *Memoirs*, *New Monthly Magazine*, p. 240. On writing and madness, see, e.g., C. Wiesenthal, *Figuring Madness in Nineteenth-Century Fiction* (Basingstoke: Macmillan, 1997).

124. See O'Connor, 'Thomas Hawkins and Geological Spectacle'.

125. T. Hawkins, *The Lost Angel and the History of the Old Adamites, Found Written on the Pillars of Seth: A Poem* (London: William Pickering, 1840). Hereafter referred to as *The Lost Angel.*

126. Anon., Review of *The Lost Angel, Metropolitan Magazine,* 29 (1840), pp. 102–3, at p. 103.

127. Ibid., p. 103.

128. Ibid., p. 103.

129. [J. Mitford], Review of *The Lost Angel, Gentleman's Magazine,* 15 (1841), pp. 169–72, at p. 170.

130. Ibid., p. 170.

131. T. Hawkins, *The Wars of Jehovah in Heaven, Earth, and Hell: In Nine Books* (London, 1844). Hereafter referred to as *The Wars of Jehovah.*

132. J. Mitford, Review of *The Wars of Jehovah, Gentleman's Magazine,* 23 (1845), pp. 516–19, at p. 516.

133. Ibid., pp. 519, 517.

134. J. Heraud, Review of *The Wars of Jehovah, Athenaeum,* 923 (1845), pp. 659–60, at p. 660.

135. Ibid., p. 659.

136. Hawkins, *Memoirs,* p. vi.

137. [Lord], Review of *Memoirs,* p. 469.

4 Eccentricity on Display

1. On exhibitions in this period, see, e.g., Altick, *The Shows of London*; P. Greenhalgh, *Ephemeral Vistas: The Expositions Universelles, Great Exhibitions and World's Fairs, 1851–1939* (Manchester: Manchester University Press, 1988); S. Qureshi, 'Living Curiosities: Human Ethnological Display, 1800–1855' (PhD thesis: University of Cambridge, 2005).

2. [W. Kinsey], 'Random Recollections of a Visit to Walton Hall', *Gentleman's Magazine,* 29 (1848), pp. 33–9, at p. 33. Biographical writings on Waterton include R. Aldington, *The Strange Life of Charles Waterton* (London: Evans Brothers, 1949); L. Barber, *The Heyday of Natural History, 1820–1870* (London: Jonathan Cape, 1980), pp. 99–110; J. Blackburn, *Charles Waterton 1782–1865: Traveller and Conservationist* (London: Vintage, 1996); V. Carroll, 'Charles Waterton', in B. Lightman (ed.), *The Dictionary of Nineteenth-Century British Scientists* (Bristol: Thoemmes Continuum, 2004); B. Edginton, *Charles Waterton: A Biography* (Cambridge: The Lutterworth Press, 1996); R. Hobson, *Charles Waterton: His Home, Habits and Handiwork* (1866; London: Whittaker, 1867); R. Irwin, *Letters of Charles Waterton of Walton Hall, near Wakefield* (London: Rockliff Publishing Corporation, 1955); Sitwell, *The English Eccentrics,* pp. 262–85; Wakefield Museum, *Charles Waterton 1782–1865: Traveller and Naturalist* (Wakefield: Wakefield Museum, 1982). Waterton prefaced each volume of his *Essays on Natural History* with an autobiographical essay; these were collected together and edited by Norman Moore, a friend of Waterton, in C. Waterton, *Essays on Natural History, Chiefly Ornithology. Edited, with a Life of the Author by N. Moore* (London: Frederick Warne & Co, 1871). Hereafter referred to as *Essays* (1871). On Waterton see also V. Carroll, 'The Natural History of Visiting: Responses to Charles Waterton and Walton Hall', *Studies in History and Philosophy of Biology and Biomedical Sciences,* 35 (2004), pp. 31–64; Carroll, 'Natural History on Display: Visiting Charles Waterton', in A. Fyfe and B. Lightman

(eds), *Science in the Marketplace: Nineteenth-Century Sites and Experiences* (Chicago, IL: University of Chicago Press, 2007), pp. 271–300.

3. Anon., 'The Home of Waterton, the Naturalist', *The Leisure Hour*, 8 (1859), pp. 486–7, at p. 486.

4. These were later collected as C. Waterton, *Essays on Natural History, Chiefly Ornithology* (London: Longman, 1st series, 1838; 2nd series, 1844; 3rd series, 1857).

5. Sitwell, *The English Eccentrics*.

6. C. Waterton, *Wanderings in South America, the North-West of the United States and the Antilles, in the Years 1812, 1816, 1820 & 1824, with Original Instructions for the Perfect Preservation of Birds, Etc. For Cabinets of Natural History* (London: Mawman, 1825). Hereafter referred to as *Wanderings*.

7. Edginton, *Charles Waterton*, p. 1.

8. See, e.g., S. Bann, 'Introduction', in S. Bann (ed.), *Frankenstein, Creation and Monstrosity* (London: Reaktion, 1994), pp. 1–15; C. Grasseni, 'Taxidermy as Rhetoric of Self-Making: Charles Waterton (1782–1865), Wandering Naturalist', *Studies in History and Philosophy of Biology and Biomedical Sciences*, 29 (1998), pp. 269–94.

9. On building, see R. Wilson and A. Mackley, *Creating Paradise: The Building of the English Country House 1660–1880* (London & New York: Hambledon, 2000). On layout, see J. Franklin, *The English Country House and Its Plan* (London and Boston, MA: Routledge & Kegan Paul, 1981). On day-to-day life, see, e.g., J. Gerard, *Country House Life: Family and Servants, 1815–1914* (Oxford and Cambridge, MA: Blackwell, 1994); M. Girouard, *Life in the English Country House: A Social and Architectural History* (New Haven, CT, and London: Yale University Press, 1978). On visiting, see P. Mandler, *The Fall and Rise of the Stately Home* (New Haven, VT, and London: Yale University Press, 1997); E. Moir, *The Discovery of Britain: The English Tourists 1540–1840* (London: Routledge & Keegan Paul, 1964); I. Ousby, *The Englishman's England: Taste, Travel and the Rise of Tourism* (Cambridge: Cambridge University Press, 1990), ch. 2; A. Tinniswood, *A History of Country House Visiting: Five Centuries of Tourism and Taste* (Oxford: Basil Blackwell and the National Trust, 1989).

10. On nineteenth-century natural history museums, see, e.g., J. Thackray and B. Press, *The Natural History Museum: Nature's Treasurehouse* (London: Natural History Museum, 2001); C. Yanni, *Nature's Museums: Victorian Science and the Architecture of Display* (London: The Athlone Press, 1999). While research has tended to centre on large-scale, public, metropolitan institutions, recently attention has been turned to provincial collections. See, e.g., S. Alberti, 'Placing Nature: Natural History Collections and Their Owners in Nineteenth-Century Provincial England', *British Journal for the History of Science*, 35 (2002), pp. 291–311. On the wider cultural relevance of collecting in the nineteenth century, see, e.g., B. Black, *On Exhibit: Victorians and Their Museums* (Charlottesville, VA, and London: University Press of Virginia, 2000). On collecting, see also, e.g., J. Elsner and R. Cardinal, *The Cultures of Collecting* (London: Reaktion, 1994); S. Pearce, *On Collecting: An Investigation into Collecting in the European Tradition* (London and New York: Routledge, 1995); K. Pomian, *Collectors and Curiosities: Paris and Venice, 1500–1800*, trans. E. Wiles-Porter (first published in French 1987; Cambridge: Polity Press, 1990); S. Stewart, *On Longing: Narratives of the Miniature, the Gigantic, the Souvenir, the Collection* (Durham, NC, and London: Duke University Press, 1993). For a wonderful response-oriented analysis of a museum which focuses on the intentions of the collector, see S. Bann, *Under the Sign: John Bargrave as Collecter, Traveler, and Witness* (Ann Arbor: University of Michigan Press, 1994).

11. Anon., 'The Home of Waterton' *The Leisure Hour*, pp. 486–7; J. Brooke, *Reminiscences of Charles Waterton, the Naturalist: From a Visit in 1861* (Wigan: Thomas Birch, 1877); J. Byrne, *Social Hours with Celebrities: Being the Third and Fourth Volumes of 'Gossip of the Century'* (London: Ward & Downey, 1898), vol. 2, pp. 32–116; T. Dibdin, *Bibliographical, Antiquarian and Picturesque Tour in the Northern Counties of England and in Scotland*, 3 vols (London, 1838); Henry Dixon, *Saddle and Sirloin: Or, English Farm and Sporting Worthies* (London: Rogerson & Tuxford, 1870), pp. 310–19; P. Fitzgerald, *Memoirs of an Author*, 2 vols (London: Richard Bentley & Son, 1894), vol. 1, pp. 225–38; G. Harley, 'Reminiscences of Charles Waterton', *Selbourne Magazine*, 2 (1889), pp. 20–3, 35–7, 114–17, 165–8; G. Head, *A Home Tour through the Manufacturing Districts of England, in the Summer of 1835* (1836; London and Edinburgh: Frank Cass & Co, 1968); Hobson, *Charles Waterton*; [Kinsey], 'Random Recollections'; S. Menteath, 'Some Account of Walton Hall, the Seat of Charles Waterton, Esq', *Magazine of Natural History*, 8 (1834), pp. 28–36, reprinted as 'Walton Hall' in *The Youth's Instructor and Guardian*, 19 (1835), pp. 92–7, 125–9; J. Wood, 'Modern Taxidermy', *Cornhill Magazine*, 1:7 (1863), pp. 120–5.
12. Harley, 'Reminiscences of Charles Waterton', passim.
13. Ibid., p. 36.
14. Edginton, *Charles Waterton*, p. 15.
15. Mandler, *The Fall and Rise of the Stately Home*, ch. 1.
16. Brooke, *Reminiscences*, p. 6.
17. Anon., 'The Home of Waterton', *The Leisure Hour,* p. 487.
18. Harley, 'Reminiscences of Charles Waterton', p. 35.
19. On Waterton's quest in search of the wourali poison, see Waterton, *Wanderings* (1825), pp. 1–76. Waterton believed that the compound might be effective in treating hydrophobia.
20. Harley, 'Reminiscences of Charles Waterton', pp. 22–3.
21. Byrne, *Social Hours with Celebrities*, vol. 2, p. 34.
22. Ibid., p. 33.
23. Harley, 'Reminiscences of Charles Waterton', p. 35.
24. Ousby, *The Englishman's England*, p. 10; Wilson and Mackley, *Creating Paradise*, p. 79.
25. Harley, 'Reminiscences of Charles Waterton', p. 35.
26. T. Dibdin, *The Bibliomania; Or Book-Madness; Containing Some Account of the History, Symptoms, and Cure of This Fatal Disease; in an Epistle Addressed to Richard Heber, Esq.* (London, 1809).
27. T. Dibdin, *A Bibliographical, Antiquarian and Picturesque Tour in France and Germany*, 3 vols (London, 1821).
28. Dibdin, *Bibliographical, Antiquarian and Picturesque Tour in the Northern Counties of England and in Scotland*, vol. 1, p. 148. This was unusual, as many country-house owners opened their houses to visitors only when they were away.
29. Dibdin, *Bibliographical, Antiquarian and Picturesque Tour in the Northern Counties of England and in Scotland*, vol. 1, p. 147.
30. Ibid., vol. 1, pp. 148–9.
31. G. Head, *Forest Scenes and Incidents in the Wilds of North America: Being a Diary of a Winter's Route from Halifax to the Canadas, and During Four Months' Residence in the Woods on the Borders of Lakes Huron and Simcoe* (London: J. Murray, 1829).
32. Head, *A Home Tour*, p. 153.
33. A. Wallace, *Walking, Literature, and English Culture: The Origins and Uses of Peripatetic in the Nineteenth Century* (Oxford: Clarendon Press, 1993), p. 64. On perceptions of rail travel, see, e.g., W. Schivelbusch, *The Railway Journey: Trains and Travel in the 19th*

Century (Oxford: Basil Blackwell, 1977). On how changing notions of travel affected attitudes towards country houses, see, e.g., Wilson and Mackley, *Creating Paradise*, ch. 3.

34. Tinniswood, *A History of Country House Visiting*, pp. 132–50.

35. C. Knight, *Knight's Excursion-Train Companion* (London: Charles Knight, 1851), p. iv.

36. [Kinsey], 'Random Recollections', p. 36.

37. The ticket in Figure 4.5 is from a collection of thirteen letters to T. Allis, York City Archives, accession no. NRA9540.

38. A great deal has been written about the development of picturesque theory during the late eighteenth and early nineteenth century, and its impact on painters, garden designers, architects, poets and travellers. On picturesque theory, see, e.g., C. Hussey, *The Picturesque: Studies in a Point of View* (1927; London: Frank Cass & Co, 1967). On the picturesque in painting, see, e.g., J. Barrell, *The Idea of Landscape and the Sense of Place*. On picturesque garden design, see, e.g, E. Hyams, *Capability Brown & Humphry Repton* (London: J. M. Dent & Sons Ltd., 1971); D. Watkin, *The English Vision: The Picturesque in Architecture, Landscape and Garden Design* (London: John Murray, 1982). On picturesque travel, see, e.g., M. Andrews, *The Search for the Picturesque: Landscape, Aesthetics and Tourism in Britain, 1760–1800* (Stanford, CA: Stanford University Press, 1989); J. Buzard, *The Beaten Track: European Tourism, Literature, and the Ways to Culture, 1800–1918* (Oxford: Clarendon Press, 1993). On the popularization of the picturesque, see, e.g., A. Bermingham, *Landscape and Ideology: The English Rustic Tradition 1740–1860* (London: Thames & Hudson, 1987). For a contemporary account, see e.g. W. Gilpin, *Three Essays: On Picturesque Beauty; on Picturesque Travel; and on Sketching Landscape: To Which Is Added a Poem, on Landscape Painting* (London: R. Bladmire, 1792).

39. Harley, 'Reminiscences of Charles Waterton', p. 35.

40. Dibdin, *Bibliographical, Antiquarian and Picturesque Tour in the Northern Counties of England and in Scotland*, vol. 1, p. 149.

41. [Kinsey], 'Random Recollections', p. 35.

42. Anon., 'The Home of Waterton', *The Leisure Hour*, p. 486.

43. *The Leisure Hour*, 8 (1859), pp. 420–22, 438–42, 461–63, 474–6.

44. Although Waterton was worth less than £14,000 upon his death in 1782 he remained, compared to most, a gentleman of wealth and fortune.

45. Harley, 'Reminiscences of Charles Waterton', p. 36.

46. Head, *Home Tour*, p. 135.

47. Byrne, *Social Hours with Celebrities*, vol. 2, pp. 43–4.

48. Ibid., p. 44.

49. Harley, 'Reminiscences of Charles Waterton', p. 165.

50. Byrne, *Social Hours with Celebrities*, vol. 2, pp. 46–7.

51. Head, *Home Tour*, p. 153; Hobson, *Charles Waterton*, p. 190.

52. Harley, 'Reminiscences of Charles Waterton', p. 115.

53. P. Farber, 'The Development of Taxidermy and the History of Ornithology', *Isis*, 68 (1977), pp. 550–66, at p. 562. See also P. Farber, *The Emergence of Ornithology as a Scientific Discipline, 1760–1850* (Dordrecht: Reidel, 1982).

54. Waterton, *Wanderings*, p. 307.

55. See, e.g., S. Lee, *Taxidermy: Or, the Art of Collecting, Preparing, and Mounting Objects of Natural History. For the Use of Museums and Travellers* (1821; London: Longman et al., 1843); the 1843 edition includes a section on Waterton's method.

56. Waterton, *Essays* (1871), pp. 519–20.

57. W. Swainson, *On the Natural History and Classification of Birds* (London: Longman, 1836), p. 211. On quinary classification, see, e.g., H. Ritvo, *The Platypus and the Mermaid*, pp. 31–5.

58. Waterton, *Essays* (1871), pp. 512, 513, 521.

59. Ibid., p. 534.

60. Ibid., pp. 534, 536.

61. Wood, 'Modern Taxidermy', p. 122.

62. Hobson, *Charles Waterton*, p. 189.

63. S. Bann, 'The Historian as Taxidermist: Ranke, Barante, Waterton', in E. S. Shaffer (ed.), *Comparative Criticism; A Yearbook*, 3 (Cambridge: Cambridge University Press, 1981), pp. 21–49. On objectivity, see L. Daston and P. Galison, 'The Image of Objectivity', *Representations*, 40 (1992), pp. 81–128. On taxidermic realism, see S. L. Star, 'Craft vs. Commodity, Mess vs. Transcendence: How the Right Tool Became the Wrong One in the Case of Taxidermy and Natural History', in A. Clarke and J. Fujimura (eds), *The Right Tools for the Job: At Work in Twentieth-Century Life Sciences* (Princeton, NJ: Princeton University Press, 1992), pp. 257–86, at p. 260.

64. Anon., 'Wanderings in South America Etc.', *Monthly Review*, 108 (1825), pp. 66–77, at p. 71.

65. Bann, 'Introduction' to *Frankenstein, Creation and Monstrosity*, p. 8.

66. Waterton, *Wanderings*, p. 308.

67. Wood, 'Modern Taxidermy', p. 124.

68. Edginton, *Chares Waterton*, p. 118.

69. Dibdin, *Bibliographical, Antiquarian and Picturesque Tour*, p. 150.

70. [J. Barrow], 'Wanderings in South America Etc.', *Quarterly Review*, 33 (1825), pp. 314–32, at p. 332.

71. [S. Smith], 'Wanderings in South America Etc.', *Edinburgh Review*, 43 (1826), pp. 299–315, at p. 307.

72. Anon., 'Wanderings in South America Etc.', *Literary Gazette and Journal of Belles Lettres* (1826), no. 468, pp. 4–8; no. 470, pp. 39–40; no. 471, pp. 56–7; at no. 468, p. 4, note.

73. Wood, 'Modern Taxidermy', p. 124.

74. Ritvo, *The Platypus and the Mermaid*, p. 55.

75. J. Browne and S. Messenger, 'Victorian Spectacle: Julia Pastrana, the Bearded and Hairy Female', *Endeavour*, 27 (2003), pp. 155–9.

76. Head, *Home Tour*, p. 162.

77. Wood, 'Modern Taxidermy', p. 124.

78. E. B. Tweedie, *George Harley Frs or the Life of a London Physician* (London: The Scientific Press, 1899), p. 278.

79. On the 'abhuman' in Gothic literature, see K. Hurley, *The Gothic Body: Sexuality, Materialism and Degeneration at the Fin De Siècle* (Cambridge: Cambridge University Press, 1996). On atavism, see, e.g., R. Mighall, *A Geography of Victorian Gothic Fiction: Mapping History's Nightmares* (Oxford: Oxford University Press, 1999), ch. 4.

80. Wood, 'Modern Taxidermy', p. 124. On mermaids, see, e.g., Ritvo, *The Platypus and the Mermaid*, pp. 178–82.

81. Brooke, *Reminiscences*, p. 10.

82. Brooke, *Reminiscences*, pp. 10–11.

83. Dixon, *Saddle and Sirloin*, pp. 315, 314.

84. Fitzgerald, *Memoirs*, vol. 1, p. 228.

85. Hobson, *Charles Waterton*, p. 211.

86. A. von Humboldt and A. Bonpland, *Voyage aux régions équinoxiales du nouveau continent, fait dans les années 1799 à 1804*, 30 vols (Paris: Schoell, 1805–34); A. von Humboldt, *Personal Narrative of Travels to the Equinoctial Regions of the New Continent*, trans. H. M. Williams (London: Longman et al., 1814–29). On Humboldt, see, e.g., M. Dettlebach, 'Humboldtian Science', in N. Jardine, J. A. Secord and E. C. Spary, *Cultures of Natural History* (Cambridge: Cambridge University Press, 1992), pp. 287–304; N. Leask, *Curiosity and the Aesthetics of Travel Writing 1770–1840* (2002; Oxford: OUP, 2004), ch 6; M. L. Pratt, *Imperial Eyes: Studies in Travel Writing and Transculturation* (London and New York: Routledge, 1992), pp. 111–43.
87. On nineteenth-century travel writing in general, see, e.g., C. Blanton, *Travel Writing: The Self and the World* (1995; New York and London: Routledge, 2002); J. Duncan and D. Gregory (eds), *Writes of Passage: Reading Travel Writing* (London and New York: Routledge, 1999); P. Hulme and T. Youngs (eds), *The Cambridge Companion to Travel Writing* (Cambridge: Cambridge University Press, 2002); B. Korte, *English Travel Writing from Pilgrimages to Postcolonial Explorations*, trans. C. Matthias (Basingstoke: Macmillan, 2000); D. Seed, 'Nineteenth-Century Travel Writing: An Introduction', *The Yearbook of English Studies (Nineteenth-Century Travel Writing)*, 34 (2004). On imperialism and nineteenth-century scientific travel writing, see, e.g., D. Miller and P. Reill (eds), *Visions of Empire: Voyages, Botany and Representations of Nature* (Cambridge: Cambridge University Press, 1996); Pratt, *Imperial Eyes*. On the aesthetics of nineteenth-century scientific travel writing, see, e.g., Leask, *Curiosity and the Aesthetics of Travel Writing 1770–1840*.
88. R. Bridges, 'Exploration and Travel Outside Europe (1720–1914)', in P. Hulme and T. Youngs (eds), *The Cambridge Companion to Travel Writing* (Cambridge: Cambridge University Press, 2002), pp. 53–69, at p. 53.
89. See, e.g., J. Clifford, *The Predicament of Culture* (Cambridge: Harvard University Press, 1988), pp. 215–51. On natural historical travel, see, e.g., N. Jardine, J. A. Secord and E. Spary (eds), *Cultures of Natural History* (Cambridge: Cambridge University Press, 1992).
90. Pratt, *Imperial Eyes*, pp. 199, 24–35.
91. Waterton, *Wanderings*, pp. 38–9.
92. J. Duncan: 'Dis-Orientation: On the Shock of the Familiar in a Far-Away Place', in Duncan and Gregory, *Writes of Passage*, pp. 151–63.
93. Head, *Home Tour*, p. 157.
94. See, e.g., A. Fyfe, 'Reading Natural History at the British Museum and the *Pictorial Museum*', in Fyfe and Lightman, *Science in the Marketplace*, pp. 196–230.
95. Head, *Home Tour*, p. 157.
96. Buzard has argued that nineteenth-century travellers in the picturesque tradition drew on 'an extended range of metaphors, taking as models not only landscape painting but also such related visual arts as drama and *tableaux vivants*': Buzard, *The Beaten Track*, p. 187.
97. Waterton, *Wanderings*, p. 227.
98. Ibid., p. 230.
99. Ibid., p. 230.
100. Ibid., p. 231.
101. [Kinsey], 'Random Recollections', p. 38.
102. Anon., 'Wanderings in South America Etc.', *Literary Gazette*, 468, pp. 4–8; [Smith], 'Wanderings in South America Etc.', pp. 310–13; [Barrow], 'Wanderings in South Amer-

ica Etc.', pp. 320–23; Anon., 'Wanderings in South America Etc.', *British Critic*, 2 (1826), pp. 93–105, at p. 103.

103. Wakefield Museum, *Charles Waterton*, p. 10.

104. Head, *Home Tour*, p. 158.

105. Menteath, 'Some Account of Walton Hall', p. 31.

106. Head, *Home Tour*, p. 157.

107. Dibdin, *Bibliographical, Antiquarian and Picturesque Tour*, vol. 1, p. 150.

108. Ibid., vol. 1, p. 146.

109. James Rennie called him 'the eccentric Waterton'. See C. Waterton, *Essays on Natural History, Chiefly Ornithology* (London: Longman, 1838), p. iv.

110. M. Luckhurst and J. Moody (eds), *Theatre and Celebrity in Britain, 1660–2000* (London: Palgrave Macmillan, 2005), p. 3. Perhaps the best-known definition of celebrity is Daniel Boorstin's 'The celebrity is a person who is known for his well-knownness.' Boorstin's definition is geared towards the study of twentieth-century, American celebrities, however; it is less useful for making sense of the relationship between celebrity and eccentricity in nineteenth-century Britain because it does not fully take account of the historicity of celebrity culture. D. Boorstin, *The Image; Or What Happened to the American Dream* (1962; London: Pelican Books, 1963), p. 67.

111. Hill, *The Fanciad*, p. 27.

112. Braudy, *The Frenzy of Renown*, p. 7.

113. Ibid., p. 13.

114. On virtuosity see, P. Metzner, *Crescendo of the Virtuoso: Spectacle, Skill, and Self-Promotion in Paris during the Age of Revolution* (Berkeley, CA: University of California Press, 1998).

115. [W. Hazlitt], 'Of Persons One Would Wish to Have Seen', *New Monthly Magazine*, 16 (1826), pp. 32–41.

116. Ibid., p. 32.

117. Ibid., p. 38.

118. Braudy, *Frenzy of Renown*, p. 444.

119. Harley, 'Reminiscences of Charles Waterton', p. 36.

120. Anon., 'The Home of Waterton', *The Leisure Hour*, p. 487.

121. Brooke, *Reminiscences*, p. 4; Harley, 'Reminiscences of Charles Waterton', p. 36.

122. Brooke, *Reminiscences*, p. 4; Hobson, *Charles Waterton*, p. 163.

123. Hobson, *Charles Waterton*, p. 252. The origins of the phrase 'What a shocking bad hat!' are traced in C. Mackay's *Memoirs of Extraordinary Popular Delusions*, 3 vols (London: R. Bentley, 1841), vol. 1, pp. 326–7. Mackay relates that there was a hotly contested election for the borough of Southwark and one of the candidates was an eminent hatter. Whilst canvassing he attempted to bribe voters without letting them realize that they were being bribed. Whenever he called on a voter whose hat was of an inferior kind he said, 'What a shocking bad hat you have got; call at my warehouse, and you shall have a new one!' On election day this was remembered: the candidate's opponents incited the crowds to chant 'What a shocking bad hat!' right through his address. For months afterwards, Mackay explains, any Londoner whose hat showed signs, however slight, of ancient service could expect to be set upon by gangs of jokers who would gleefully cry, 'What a shocking bad hat!' and convey the offending article to the gutter for the amusement of the crowds.

124. Tweedie, *George Harley FRS*, p. 268n.

125. Brooke, *Reminiscences*, p. 14.

126. Byrne, *Social Hours with Celebrities*, vol. 2, p. 53.

127. Ibid., pp. 73–5.

128. Quoted in S. Gill, *Wordsworth and the Victorians* (Oxford: Clarendon Press, 1998), p. 12.

129. Anon., 'The Home of Charles Waterton', *The Leisure Hour*, pp. 486–7.

130. Hobson, *Charles Waterton*, p. 246.

131. [Barrow], 'Wanderings in South America Etc.', pp. 319, 323.

132. [Smith], 'Wanderings in South America Etc.', p. 315.

133. See, e.g., M. B. Campbell, *The Witness and the Other World: Exotic European Travel Writing, 400–1600* (Ithaca, NY: Cornell University Press), pp. 136–48; Korte, *English Travel Writing*, pp. 19–24.

134. *The Travels of Sir John Mandeville*, trans. C. Moseley (Harmondsworth: Penguin, 1983), p. 137.

135. [H. Southern], 'Waterton's Wanderings in South America', *London Magazine*, 4 (1826), pp. 343–53, at p. 343.

136. [Rudolph Erich Raspe], *Surprising Adventures of the Renowned Baron Munchausen, Containing Singular Travels, Campaigns, Voyages, and Adventures. Also, an Account of a Voyage to the Moon and Dog Star* (first published in English 1785; London: Printed for Thomas Tegg, 1811), p. 6.

137. Ibid., pp. 3–6.

138. Ibid., pp. 4–5.

139. Ibid., p. 6.

140. J. Bruce, *Travels to Discover the Source of the Nile in the Years 1768, 1769, 1770, 1771, 1772 and 1773*, 5 vols (Edinburgh: J. Ruthven; London: G. G. J. & J. Robinson, 1790).

141. Leask, *Curiosity and the Aesthetics of Travel Writing*, pp. 54–5.

142. J. Bruce, *Travels Through Part of Africa, Syria, Egypt, and Arabia, into Abyssinia, to Discover the Source of the Nile, Performed Between the Years 1768 and 1773* (Glasgow: Printed by W. Lang, 1819), pp. 254–7.

143. Anon., 'Wanderings in South America Etc.', *Literary Gazette*, 70, p. 39.

144. Smiles, *Self Help*, p. 314, cited in Secord, 'Corresponding Interests', pp. 383–408, at p. 390.

145. Harley, 'Reminiscences of Charles Waterton', p. 22.

146. Ibid., p. 37.

147. Anon., 'Waterton, the Wanderer', *Chambers's Edinburgh Journal*, 492 (1841), pp. 188–9, at p. 188.

148. Ibid., p. 188.

149. Ibid., p. 188; Menteath, 'Walton Hall', p. 31.

150. Anon. 'Charles Waterton the Naturalist', *The Leisure Hour*, 1 (1852), pp. 282–6, at p. 283.

151. Dibdin, *Biographical, Antiquarian and Picturesque Tour*, vol. 1, pp. 147, 151.

152. On anti-tourism in the nineteenth century, see Buzard, *The Beaten Track*.

153. Byrne, *Social Hours with Celebrities*, vol. 2, pp. 76–7.

154. Anon., 'The Home of Waterton', *The Leisure Hour*, p. 487.

Conclusion

1. Cited in J. Uglow, *Elizabeth Gaskell: A Habit of Stories* (London: Faber & Faber, 1993), p. 406. On anecdotes, see, e.g., J. Fineman, 'The History of the Anecdote: Fiction and Fiction', in H. Veser (ed.), *The New Historicism* (New York: Routledge, 1989), pp. 49–76; L. Gossman, 'Anecdote and History', *History and Theory*, 42 (2003), pp. 143–68. Existing discussions of the anecdote tend to focus on the role of the anecdote in the writing of history, rather than the history of the anecdote itself as a cultural form.

2. Anon., 'Wanderings in South America Etc.', *Monthly Review*, p. 66.

3. Anon., 'Essays on Natural History', *Dublin Review*, 43 (1857), p. 532.

4. Anon., 'Wanderings in South America Etc.' *Literary Gazette*, 468, pp. 4–5.

5. Anon., Review of *Memoirs*, *New Monthly Magazine*, p. 240.

6. Campbell, *The Witness and the Other World*, p. 6; Korte, *English Travel Writing*, p. 9.

7. Martin, *The Total Defeat of the Yearly Meeting of the British Association of Scientific Asses*.

8. See, e.g., J. H. Brooke, *Science and Religion: Some Historical Perspectives* (Cambridge; Cambridge University Press, 1991); Lindberg and Numbers, *God and Nature*.

9. Lyell in a lecture delivered at King's College London, 4 May 1832; cited in Moore, 'Geologists and Interpreters of Genesis in the Nineteenth Century', p. 337. On the professionalization of geology, see, e.g., Porter, 'Gentlemen and Geology'.

10. *Illustrated London News* (17 June 1865), p. 583.

11. Stephen Bann similarly interprets the modest collection of John Bargrave, seventeenth-century traveller, virtuoso and descendant of 'the Fallen House of Bargrave', at Canterbury Cathedral as a work of mourning for the earlier losses to which his ancestors were subjected: Bann, *Under the Sign*, p. 65. On the museum as mausoleum, see J. Siegel, *Desire and Excess: The Nineteenth-Century Culture of Art* (Princeton, NJ, and Oxford: Princeton University Press, 2000). On the Dulwich Picture Gallery, designed as a museum-cum-mausoleum by John Soane, see D. Watkin, 'Monuments and Mausolea in the Age of Enlightenment', in G. Waterfield (ed.), *Soane and Death: The Tombs and Monuments of Sir John Soane* (London: Dulwich Picture Gallery, 1996), pp. 9–25.

12. Hobson, *Charles Waterton*, p. 256.

13. Wood, 'Modern Taxidermy', pp. 123–4.

14. www.bw-watertonparkhotel.co.uk (accessed 24 April 2008).

15. D. Bellamy, 'Introduction', in Charles Waterton, *Wanderings in South America* (London: Century Publishing Co, 1984), pp. xvii–xix; G. Durrell, 'Foreword', in Blackburn, *Charles Waterton*, p. ix (foreword dated 1988); Edginton, *Charles Waterton*, p. ix.

16. Keay, *Eccentric Travellers*, pp. 97–125; J. Joliffe, *Eccentrics* (London: Duckworth, 2001), pp. 17–22; Timpson, *Timpson's English Eccentrics*, pp. 14–17; Reader's Digest, *Great British Eccentrics*, pp. 37–8.

17. See, e.g., F. Taylor, 'Yorkshire's Naturalist', *The Dalesman*, 19 (1957), pp. 13–14; J. West, 'The Squire Brought 'Em Back Alive', *Yorkshire Life Illustrated* (September 1958), pp. 22–3; B. Lonsdale, 'A Fantastic Squire at Walton Hall', *The Dalesman*, 27 (1965), pp. 120–2; H. Atha, 'Exploits of the Eccentric Squire', *Yorkshire Post Colour Magazine* (5 June 1982), pp. 20–1; John Nettleship, 'A Yorkshire Naturalist's Debt to Lancashire', *Yorkshire Ridings Magazine*, 19 (1982), p. 77; R. Bell, 'Waterton's World of Wildlife', *Yorkshire Life* (June 1982), pp. 14–15; C. Ritchie, 'A Prophet at Walton Hall', *The Dalesman*, 44 (1983), pp. 763–5; Anon., 'Yorkshire Curiosities: An Eccentric Squire', *The Dalesman*, 54 (1993), p. 53, adapted from J. D. Smith and T. Smith, *South and West Yorkshire Curiosities* (Wimborne: Dovecote Press, 1992). On Waterton as a Yorkshire

character, see, e.g., W. Smith (ed.), *Old Yorkshire* (London: Longmans, Green, 1882), pp. 120–6; M. Colbeck, *Queer Folk: A Comicality of Yorkshire Characters* (Manchester: The Whitehorn Press, 1977), pp. 30–9; L. Cooper, *Great Men of Yorkshire (West Riding)* (London: The Bodley Head, 1955), vol. 3, pp. 62–80.

18. www.wakefield.gov.uk/CultureAndLeisure/Museums/Wakefield/default.htm (accessed 24 April 2008).

19. C. Lever, 'Father of Wildlife Conservation', *Country Life*, 171 (1982), pp. 1698–9. See also G. Christian, 'A Naturalist in Advance of His Time', *Country Life*, 137 (1965), pp. 1290–1.

20. Wakefield Museum, *Charles Waterton 1782–1865*, p. 2.

21. Blackburn, *Charles Waterton*, printed cover.

22. Welford, *Men of Mark 'Twixt Tyne and Tweed*, vol. 1, p. 2.

23. Ibid., vol. 3, p. 175.

24. Anon., 'Northern Profile', *The Nor'-Easter Staff Magazine of the North Eastern Electricity Board and Central Electricity Generating Board, North Eastern Region, Northern Division*, 7 (1966), p. 10.

25. E. Forster, 'Just a Moment', *Evening Chronicle* (23 November 1972), p. 18.

26. Anon., 'Hat's Off to William', *Evening Chronicle* (18 August 1987), p. 6.

27. J. Latimer, *Local Records; Or, Historical Register of Remarkable Events, Which Have Occurred in Northumberland & Durham, Newcastle-Upon-Tyne, and Berwick-Upon-Tweed etc.* (Newcastle: The Chronicle Office, 1857), p. 291.

28. G.B.R., 'Billy Martin', *Newcastle Courant* (31 January 1879), p. 6; Scott, *Autobiographical Notes of the Life of William Bell* Scott, vol. 1, p. 197. Scott's account was excerpted into Welford's *Men of Mark 'Twixt Tyne and Tweed* and thus effectively became common knowledge. Due to a typographical error, however, the remark which might have cast doubts on the accuracy of Latimer's description of Martin's hat was erased from record: Welford's version reads nonsensically, 'we encountered the well-known figure in his extraordinary skull-cap, decorated with military surtout closely buttoned to the throat': Welford, *Men of Mark 'Twixt Tyne and Tweed*, vol. 3, p. 177.

29. R. Hewison, *The Heritage Industry: Britain in a Climate of Decline* (London: Methuen, 1987), p. 9.

30. Ibid., p. 10.

31. D. Lowenthal, *The Past is a Foreign Country* (Cambridge: Cambridge University Press, 1985); R. Samuel, *Theatres of Memory. Volume 1: Past and Present in Contemporary Culture* (London: Verso, 1994). See also P. Wright, *On Living in an Old Country: The National Past in Contemporary Britain* (London: Verso, 1985).

32. Lyrics by Gary Miller of The Whisky Priests, provided by Keith Armstrong in personal correspondence, 17 November 2003. A review of the performance by 'H. S.' was placed on the web shortly afterwards at www.swaddleh.freeserve.co.uk/today/mad.htm (accessed 24 April 2008).

33. Personal correspondence, 5 April 2005.

34. Hawkins also continues to be of interest to Somerset antiquarians, and accounts of his life and collecting practices are to be found in the *Proceedings of the Somersetshire Archaeological & Natural History Society*. See, e.g., A. Bulleid, A., 'Notes on the Life and Work of Thomas Hawkins F.G.S., pp. 59–71; Taylor, 'Joseph Clark III's Reminiscences about the Somerset Fossil Reptile Collector Thomas Hawkins (1810–1889)', pp. 1–10.

35. Thackray and Press, *The Natural History Museum*, p. 35.

36. On the polemical uses of history by practitioners of the sciences, see N. Jardine, *The Scenes of Inquiry: On the Reality of Questions in the Sciences* (Oxford: Oxford University Press, 2000), ch. 6.

37. McGowan, *The Dragon Seekers*; R. Forrest, The Plesiosaur Site: www.plesiosaur.com (accessed 2004).

38. Purcell and Gould, *Finders, Keepers*, p. 95; Taylor, 'Thomas Hawkins', *ODNB*; Purcell and Gould, *Finders Keepers*, p. 105.

39. Owen, *The Life of Richard Owen*, vol. 1, pp. 165–6.

WORKS CITED

A Disciple, *Martin's Philosophy!* (Newcastle: Edward Walker, 1828), in *Billy Martin Philosopher's Nonsense*.

—, *A New Song, Called the 'Northumberland Eagle Mail'* (*c.* 1828), in *Billy Martin Philosopher's Nonsense*.

Abrams, M. H., *The Mirror and the Lamp: Romantic Theory and the Critical Tradition* (1953; Oxford: Oxford University Press, 1971).

Adams, M., 'John Martin and the Prometheans', *History Today*, 56:8 (2006), pp. 40–7.

Alberti, S., 'Placing Nature: Natural History Collections and Their Owners in Nineteenth-Century Provincial England', *British Journal for the History of Science*, 35 (2002), pp. 291–311.

Aldington, R., *The Strange Life of Charles Waterton* (London: Evans Brothers, 1949).

Altick, R., *Lives and Letters: A History of Literary Biography in England and America* (New York: Alfred A Knopf, 1966).

—, *The Shows of London* (Cambridge, MA: The Belknap Press of Harvard University Press, 1978).

—, *The English Common Reader: A Social History of the Mass Reading Public, 1800–1900* (1957; Columbus: Ohio State University Press, 1998).

Amigoni, D. (ed.), *Life Writing and Victorian Culture* (Aldershot: Ashgate, 2006).

Anderson, K., 'The Weather Prophets: Science and Reputation in Victorian Meteorology', *History of Science*, 37 (1999), pp. 178–216.

Andrews, M., *The Search for the Picturesque: Landscape, Aesthetics and Tourism in Britain, 1760–1800* (Stanford, CA: Stanford University Press, 1989).

Ang, I., *Desperately Seeking the Audience* (London: Routledge, 1991).

Anon., *Shaftsbury's Farewel: Or, the New Association* (London: Printed by Walter Davis, 1683).

—, 'Old Bailey', *The Times* (18 July 1787), p. 3.

—, *Eccentric Biography; Or, Sketches of Remarkable Characters, Ancient and Modern, Including Potentates, Statesmen, Divines, Historians, Naval and Military Heroes, Philosophers, Lawyers, Impostors, Poets, Painters, Players, Dramatic Writers, Misers, &C. &C. &C. The Whole Alphabetically Arranged; and Forming a Pleasing Delineation of the Singularity,*

Whim, Folly, Caprice, &C. &C. Of the Human Mind. Ornamented with Portraits of the Most Singular Characters Noticed in the Work (London: Vernor & Hood, 1801).

—, *Eccentric Biography: Or, Memoirs of Remarkable Female Characters, Ancient and Modern. Including Actresses, Adventurers, Authoresses, Fortune-Tellers, Gipsies, Dwarfs, Swindlers, Vagrants, and Others Who Have Distinguished Themselves by Their Chastity, Dissipation, Intrepidity, Learning, Abstinence, Credulity, &C. &C. Alphabetically Arranged. Forming a Pleasing Mirror of Reflection to the Female Mind. Ornamented with Portraits of the Most Singular Characters in the Work* (London: J. Cundee, 1803).

—, *The New Cries of London, with Characteristic Engravings* (London: Printed and sold by Darton & Harven, 1804).

—, *The Cabinet of Curiosities: Or Mirror of Entertainment. Being a Selection of Extraordinary Legends; Original and Singularly Curious Letters; Whimsical Inscriptions; Ludirous Bills; Brilliant Bon Mots; Ingenious Calculations; Witty Peititions, and a Variety of Other Eccentric Matter* (London: Burkett & Plumpton, *c.* 1810).

—, *The Eccentric Magazine; Or Lives and Portraits of Remarkable Characters* (London: G. Smeeton, 1812).

—, 'William Martin, the Natural Philosopher!', *The Newcastle Magazine*, 1:1 (1822), pp. 25–9.

—, 'Untitled', *The Times* (27 January 1823), p. 2.

—, 'Wanderings in South America Etc.', *Monthly Review*, 108 (1825), pp. 66–77.

—, 'Commission of Lunacy', *The Times* (8 August 1825), p. 2.

—, *Eccentric Biography; Or Lives of Extraordinary Characters; Whether Remarkable for Their Splendid Talents, Singular Propensities, or Wonderful Adventures* (London: Thomas Tegg, 1826).

—, 'Wanderings in South America Etc.', *Literary Gazette and Journal of Belles Lettres* (1826), no. 468, pp. 4–8; no. 470, pp. 39–40; no. 471, pp. 56–7.

—, 'Wanderings in South America Etc.', *British Critic*, 2 (1826), pp. 93–105.

—, 'Prerogative Court', *The Times* (13 April 1826), p. 3.

—, 'Astronomical Lectures', *Tyne Mercury* (6 March 1827), p. 1.

—, 'To Be Sold by Auction at Mr Fletcher's, the Turk's Head Inn, Bigg Market, Newcastle ... A Newly-Built Dwelling House', *Newcastle Courant* (8 September 1827), p. 1.

—, *Imperial Mandate* (Newcastle: Blagburn, 1827).

—, *True Philosophy!* (1827).

—, *The Martin and the Eagle*, Offprint of article from the *Durham County Advertiser* (11 October 1828).

—, 'From the *Tyne Mercury*', Offprint of article in the *Tyne Mercury*, 6 October (Newcastle upon Tyne: William Mitchell, 1828).

—, 'Mechanical and Optical Exhibition, at Mr Fletcher's Long Room, Turk's Head, Bigg Market', *Tyne Mercury* (1 April 1828), p. 3.

—, *Martin Victorious!* (Newcastle: Edward Walker, 1828).

—, Untitled Advertisement for Mr Lloyd's lecture, *Tyne Mercury* (27 May 1828), p. 3.

—, 'Mr Walker's Lecture on Astronomy', *Tyne Mercury* (27 May 1828), p. 3.

—, 'Now Exhibiting at the Turks Head Long Room, Sinclair & Co's Grand Dioramic or Peristrephic New Panorama of the Battle of Navarino', *Newcastle Courant* (28 June 1828), p. 1.

—, 'Queens Head Inn, Morpeth. A Meeting of the Committee Assembled for the Adoption of Measures for Building a New Bridge at Morpeth', *Newcastle Courant* (20 September 1828), p. 1.

—, *Triumph of Truth* (North Shields: Pollock, 1828).

—, *Astronomy* (1828).

—, 'Police', *The Times* (18 February 1830), p. 3.

—, 'Law Report', *The Times* (8 May 1830), p. 4.

—, 'Mineral Kingdom', *The Penny Magazine of the Society for the Diffusion of Useful Knowledge*, 2 (1833), pp. 347–9.

—, 'Organic Remains Restored', *Penny Magazine*, 2 (1833), pp. 409–10.

—, 'Memoirs of Ichthyosauri and Plesiosauri, Etc.', *The Literary Gazette and Journal of Belles Lettres*, 910 (1834), pp. 444–5.

—, 'Memoirs of Ichthyosauri and Plesiosauri', *New Monthly Magazine* (1834), pp. 240–2.

—, 'Memoirs of Ichthyosauri and Plesiosauri. By Thomas Hawkins', *Metropolitan Magazine*, 11 (1834), Appendix pp. 1–2.

—, 'Hawkins, T., F.G.S.: Memoirs of Ichthyosauri and Plesiosauri Etc.', *Magazine of Natural History*, 7 (1834), pp. 476–9.

—, 'The Book of the Great Sea-Dragons ', *New Monthly Magazine*, 59 (1840), pp. 429–31.

—, 'The Book of the Great Sea-Dragons, Ichthisauri, and Pleriosuri [*sic*] Etc.', *Metropolitan Magazine*, 28 (1840), pp. 114–15.

—, 'The Lost Angel and the History of the Old Adamites, Found Written on the Pillars of Seth', *Metropolitan Magazine*, 29 (1840), pp. 102–3.

—, 'From the London Gazettes', *The Tablet*, 1 (1840).

—, 'Waterton, the Wanderer', *Chambers's Edinburgh Journal*, 492 (1841), pp. 188–9.

—, 'William Langstaff Weddall', *Gentleman's Magazine* (1851), p. 446.

—, 'Charles Waterton the Naturalist', *The Leisure Hour*, 1 (1852), pp. 282–6.

—, 'Essays on Natural History', *Dublin Review*, 43 (1857), p. 532.

—, 'The Home of Waterton, the Naturalist', *The Leisure Hour*, 8 (1859), pp. 486–7.

—, 'The Home of Waterton, the Naturalist', *The Wakefield Express* (27 August 1859).

—, 'Industrial Biography', *The Times* (28 December 1863), p. 5.

—, *The Book of Wonderful Characters; Memoirs and Anecdotes of Remarkable and Eccentric Persons in All Ages and Countries. Chiefly from the Text of Henry Wilson and James Caulfield* (London: John Camden Hotten, *c.* 1869).

—, 'William Martin, the Anti-Newtonian Philosopher', *Monthly Chronicle of North-Country Lore and Legend*, 1 (1887), pp. 343–9.

—, 'Northern Profile', *The Nor'-Easter Staff Magazine of the North Eastern Electricity Board and Central Electricity Generating Board, North Eastern Region, Northern Division*, 7 (1966), p. 10.

—, 'Hat's Off to William', *Evening Chronicle* (18 August 1987), p. 6.

—, 'Yorkshire Curiosities: An Eccentric Squire', *The Dalesman*, 54 (1993), p. 53.

Ashe, T., *Memoirs of Mammoth: And Various Other Extraordinary and Stupendous Bones, of Incognita, or Non-Descript Animals: Found in the Vicinity of the Ohio, Etc.* (Liverpool: Printed by G. F. Harris, 1806).

Aston, N., 'From Personality to Party: The Creation and Transmission of Hutchinsonianism, c. 1725–1750', *Studies in History and Philosophy of Science*, 35 (2004), pp. 625–44.

Atha, H., 'Exploits of the Eccentric Squire', *Yorkshire Post Colour Magazine* (5 June 1982), pp. 20–1.

Babbage, C., *Reflections on the Decline of Science in England, and on Some of its Causes* (London: B. Fellowes, 1830).

—, *Passages from the Life of a Philosopher* (London: Longman, 1864).

Babcock, B., 'Introduction', in B. Babcock (ed.), *The Reversible World* (Ithaca, NY: Cornell University Press, 1978), pp. 13–33.

Bailey Langhorne, J., 'W. Martin, the Natural Philosopher', *Notes and Queries*, 4:12 (1873), p. 134.

Bakhtin, M., *Rabelais and His World*, trans. H. Iswolsky (first published in Russian 1965; Cambridge, MA: Massachusetts Institute of Technology, 1968).

Balston, T., *The Life of Jonathan Martin, Incendiary of York Minster, with Some Account of William and Richard Martin* (London: Macmillan & Co. Ltd, 1945).

Bann, S., 'The Historian as Taxidermist: Ranke, Barante, Waterton', in E. S. Shaffer (ed.), *Comparative Criticism; A Yearbook*, 3 (Cambridge: Cambridge University Press, 1981), pp. 21–49.

—, 'Introduction', in S. Bann (ed.), *Frankenstein, Creation and Monstrosity* (London: Reaktion, 1994), pp. 1–15.

—, *Under the Sign: John Bargrave as Collector, Traveler, and Witness* (Ann Arbor, MI: University of Michigan Press, 1994).

Barber, L., *The Heyday of Natural History, 1820–1870* (London: Jonathan Cape, 1980).

Barrell, J., *The Idea of Landscape and the Sense of Place 1730–1840: An Approach to the Poetry of John Clare* (Cambridge: Cambridge University Press, 1972).

[Barrow, J.], 'Wanderings in South America Etc.', *Quarterly Review*, 33 (1825), pp. 314–32.

Barthes, R., *Writing Degree Zero*, trans. A. Lavers and C. Smith (London: Cape, 1967).

Basalla, G., 'The Spread of Western Science', *Science*, 156 (1967), pp. 611–22.

Bate, J., *Romantic Ecology: Wordsworth and the Environmental Tradition* (London: Routledge, 1991).

Bazerman, C., *Shaping Written Knowledge: The Genre and Activity of the Experimental Article in Science* (Wisconsin: The University of Wisconsin Press, 1988).

Bell, R., 'Waterton's World of Wildlife', *Yorkshire Life* (June 1982), pp. 14–15.

Bellamy, D., 'Introduction', in C. Waterton, *Wanderings in South America* (London: Century Publishing Co., 1984), pp. xvii–xix.

Benedict, B., *Curiosity: A Cultural History of Early Modern Inquiry* (Chicago, IL, and London: University of Chicago Press, 2001).

Benjamin, A., G. Cantor et al. (eds), *The Figural and the Literal: Problems of Language in the History of Science and Philosophy, 1630–1800* (Manchester: Manchester University Press, 1987).

Bermingham, A., *Landscape and Ideology: The English Rustic Tradition 1740–1860* (London: Thames & Hudson, 1987).

Billy Martin Philosopher's Nonsense, Cambridge University Library, call no: 8000.c.177.

Black, B., *On Exhibit: Victorians and Their Museums* (Charlottesville, VA, and London: University Press of Virginia, 2000).

Blackburn, J., *Charles Waterton 1782–1865: Traveller and Conservationist* (London: Vintage, 1996).

Blanton, C., *Travel Writing: The Self and the World* (1995; New York and London: Routledge, 2002).

Bogdan, R., *Freak Show: Presenting Human Oddities for Amusement and Profit* (Chicago, IL, and London: University of Chicago Press, 1988).

Boorstin, D., *The Image; Or What Happened to the American Dream* (1962; London: Pelican Books, 1963).

Booth, A., 'Men and Women of the Time: Victorian Prosopographies', in D. Amigoni (ed.), *Life Writing and Victorian Culture* (Aldershot: Ashgate, 2006), pp. 41–66.

Booth, W., *The Rhetoric of Fiction* (Chicago, IL: University of Chicago Press, 1961).

Bowler, P., and I. R. Morus, *Making Modern Science: A Historical Survey* (Chicago, IL, and London: Chicago University Press, 2005).

Braudy, L., *The Frenzy of Renown: Fame and its History* (Oxford: Oxford University Press, 1986).

[Brewster, D.], 'Decline of Science', *Quarterly Review*, 43 (1830), pp. 305–42.

Bridges, R., 'Exploration and Travel Outside Europe (1720–1914)', in P. Hulme and T. Youngs (eds), *The Cambridge Companion to Travel Writing* (Cambridge: Cambridge University Press, 2002), pp. 53–69.

Brontë, C., *Shirley. A Tale*, 3 vols (London: Smith, Elder & Co, 1849).

Brooke, J., *Reminiscences of Charles Waterton, the Naturalist: From a Visit in 1861* (Wigan: Thomas Birch, 1877).

Brooke, J. H., *Science and Religion: Some Historical Perspectives* (Cambridge: Cambridge University Press, 1991).

Brooke, J., and G. Cantor, *Reconstructing Nature* (Edinburgh: T&T Clark, 1998).

Brothers, R., *A Revealed Knowledge of the Prophecies and Times. Book the First. Wrote under the Direction of the Lord God, and Published by His Sacred Command; It Being the First Sign of Warning, for the Benefit of All Nations. Containing, with Other Remarkable Things, Not Revealed to Any Other Person on Earth, the Restoration of the Hebrews to Jerusalem, by the Year of 1798, under Their Revealed Prince and Prophet, Richard Brothers*, 2 vols (London, 1794).

Browne, J., *Charles Darwin: The Power of Place* (New York: Alfred A. Knopf, 2002).

Browne, J., and Messenger, S., 'Victorian Spectacle: Julia Pastrana, the Bearded and Hairy Female', *Endeavour*, 27 (2003), pp. 155–9.

Bruce, J., *Travels to Discover the Source of the Nile in the Years 1768, 1769, 1770, 1771, 1772 and 1773*, 5 vols (Edinburgh: J. Ruthven; London: G. G. J. & J. Robinson, 1790).

—, *Travels through Part of Africa, Syria, Egypt, and Arabia, into Abyssinia, to Discover the Source of the Nile, Performed Between the Years 1768 and 1773* (Glasgow: Printed by W. Lang, 1819).

Buckland, W., *Vindiciae geologicae: Or, the Connexion of Geology with Religion, Explained in an Inaugural Lecture Delivered before the University of Oxford, May 15, 1819, on the Endowment of a Readership in Geology by His Royal Highness the Prince Regent* (Oxford: University Press, 1820).

—, *Reliquiae diluvianea: Or, Observations on the Organic Remains Contained in Caves, Fissures, and Diluvial Gravel, and on Other Geological Phenomena, Attesting the Action of an Universal Deluge* (London: John Murray, 1823).

—, 'Notice on the Megalosaurus or Great Fossil Lizard of Stonesfield', *Transactions of the Geological Society of London*, 2:1 (1824), pp. 390–6.

—, *Geology and Mineralogy Considered with Reference to Natural Theology*, 2 vols (London: William Pickering, 1836).

Bugg, G., *Scriptural Geology; Or, Geological Phenomena Consistent Only with the Literal Interpretation of the Sacred Scriptures, Upon the Subjects of the Creation and Deluge* (London: Hatchard & Son, 1826–7).

Bulleid, A., 'Notes on the Life and Work of Thomas Hawkins F.G.S.', *Somerset Archaeology and Natural History*, 89 (1943), pp. 59–71.

Bulwer Lytton, E., *Pelham; Or, the Adventures of a Gentleman*, 3 vols (London, 1828)

Burke, P., *Varieties of Cultural History* (Cambridge: Polity Press, 1997).

Burney, F., *Evelina, or, a Young Lady's Entrance into the World*, 3 vols (London: T. Lowndes, 1778).

—, *Cecilia, or Memoirs of an Heiress*, 5 vols (London: T. Payne & Son, 1782)

—, *Camilla*, 5 vols (London: T. Payne et al., 1796).

—, *The Wanderer; Or, Female Difficulties*, 5 vols (London: Longman et al., 1814).

Bushaway, B., *By Rite: Custom, Ceremony and Community in England 1700–1880* (London: Junction Books, 1982).

Bushell, P., *Great Eccentrics* (London: George Allen & Unwin, 1984).

Buzard, J., *The Beaten Track: European Tourism, Literature, and the Ways to Culture, 1800–1918* (Oxford: Clarendon Press, 1993).

Byrne, J., *Social Hours with Celebrities: Being the Third and Fourth Volumes of 'Gossip of the Century'* (London: Ward & Downey, 1898).

Cadbury, D., *The Dinosaur Hunters: The Story of Scientific Rivalry and the Discovery of the Prehistoric World* (London: Fourth Estate, 2000).

Cage, J., *Colour and Culture: Practice and Meaning from Antiquity to Abstraction* (Thames & Hudson, 1993).

Cahan, D., 'Institutions and Communities', in D. Cahan (ed.), *From Natural Philosophy to the Sciences: Writing the History of Nineteenth-Century Science* (Chicago, IL, and London: University of Chicago Press, 2003).

Campbell, M. B., *The Witness and the Other World: Exotic European Travel Writing, 400–1600* (Ithaca, NY, and London: Cornell University Press, 1988).

Cannon, S. F., *Science and Culture: The Early Victorian Period* (New York: Dawson & Science History Publications, 1978).

Carlyle, T., 'Jean Paul F. Richter', *Edinburgh Review*, 46 (1827), pp. 176–95.

—, *On Heroes, Hero-Worship and the Heroic in History: Six Lectures. Reported with Emendations and Additions* (London: James Fraser, 1841).

Carroll, V., 'Charles Waterton', in B. Lightman (ed.), *The Dictionary of Nineteenth-Century British Scientists* (Bristol: Thoemmes Continuum, 2004).

—, 'The Natural History of Visiting: Responses to Charles Waterton and Walton Hall', *Studies in History and Philosophy of Biology and Biomedical Sciences*, 35 (2004), pp. 31–64.

—, 'Natural History on Display: Visiting Charles Waterton', in A. Fyfe and B. Lightman (eds), *Science in the Marketplace: Nineteenth-Century Sites and Experiences* (Chicago, IL: University of Chicago Press, 2007), pp. 271–300.

—, '"Beyond the Pale of Ordinary Criticism": Eccentricity and the Fossil Books of Thomas Hawkins', *Isis*, 98 (2007), pp. 225–65.

Caulfield, J., *Blackguardiana; Or, a Dictionary of Rogues, Bawds, Pimps, Whores ... Illustrated with Eighteen Portraits of the Most Remarkable Professors in Every Species of Villany* (Bagshot: John Shepherd, 1795).

—, *Portraits, Memoirs, and Characters, of Remarkable Persons, from the Reign of Edward the Third, to the Revolution. Collected from the Most Authentic Accounts Extant. A New Edition, Completing the Twelfth Class of Granger's Biographical History of England; with Many Additional Rare Portraits*, 3 vols (1794; London: R. S. Kirby, 1813).

Challinor, J., *The History of British Geology: A Bibliographical Study* (Newton Abbot: David & Charles, 1971).

Chandler, J., *England in 1819: The Politics of Literary Culture and the Case of Romantic Historicism* (Chicago, IL, and London: University of Chicago Press, 1998).

Chartier, R., *The Order of Books* (Cambridge: Polity, 1994).

Chase, M., *'The People's Farm': English Radical Agrarianism 1775–1840* (Oxford: Clarendon Press, 1988).

[Chorley, H. F.], 'Remarkable and Eccentric Characters, with Numerous Illustrations', *Athenaeum*, 1111 (1849), p. 141.

Christian, G., 'A Naturalist in Advance of His Time', *Country Life*, 137 (1965), pp. 1290–1.

Clare, J., *The Village Minstrel and Other Poems* (London: Taylor & Hessey, 1821).

Clifford, J., *The Predicament of Culture* (Cambridge: Harvard University Press, 1988).

Cock, R., 'Scientific Servicemen in the Royal Navy and the Professionalisation of Science, 1816–55', in D. Knight and M. Eddy (eds), *Science and Beliefs: From Natural Philosophy to Natural Science, 1700–1900* (Aldershot: Ashgate, 2005), pp. 95–111.

Colbeck, M., *Queer Folk: A Comicality of Yorkshire Characters* (Manchester: The Whitehorn Press, 1977).

Collini, S., *Public Moralists: Political Thought and Intellectual Life in Britain 1850–1930* (Oxford: Clarendon Press, 1991).

Coney, M., 'The Implied Author in Technical Discourse', *Journal of Advanced Composition*, 5 (1988), pp. 163–72.

Cooper, L., *Great Men of Yorkshire (West Riding)* (London: The Bodley Head, 1955).

Cooter, R., *The Cultural Meaning of Popular Science: Phrenology and the Organisation of Consent in Nineteenth-Century Britain* (Cambridge: Cambridge University Press, 1984).

Copleston, E., *Advice to a Young Reviewer, with a Specimen of the Art* (Oxford, 1807).

Corsi, P., *Science and Religion: Baden Powell and the Anglican Debate, 1800–1860* (Cambridge: Cambridge University Press, 1988).

Couper, G. W., *Original Poetry* (North Shields: R. Henderson, 1828).

Cowlishaw, B., 'A Genealogy of Eccentricity' (PhD dissertation, University of Oklahoma, 1998).

Cunningham, A., and N. Jardine (eds), *Romanticism and the Sciences* (Cambridge: Cambridge University Press, 1990).

Curry, P., *A Confusion of Prophets: Victorian and Edwardian Astrology* (London: Collins & Brown, 1992).

Curwen, E. C. (ed.), *The Journal of Gideon Mantell Surgeon and Geologist Covering the Years 1818–1852* (London: Oxford University Press, 1940).

Cuvier, G., *Recherches sur les ossemens fossiles, ou l'on r'établit les caractères de plusieurs animaux dont les révolutions du globe ont d'étruit les espèces*, 5 vols (Paris: G. Dufour & E. D'Ocagne, 1821–4).

Damrosch, D., *The Narrative Covenant: Transformations of Genre in the Growth of Biblical Literature* (Cambridge, MA: Harper & Row, 1987).

Darcy, C., 'The Gentleman's Magazine', in A. Sullivan (ed.), *British Literary Magazines: The Romantic Age 1789–1836* (Westport, CT, and London: Greenwood Press, 1983), pp. 136–40.

Darnton, R., 'What Is the History of Books?' in R. Darnton (ed.), *The Kiss of Lamourette: Reflections in Cultural History* (London: Faber & Faber, 1990), ch. 1.

Daston, L., and P. Galison, 'The Image of Objectivity', *Representations*, 40 (1992), pp. 81–128.

Davidoff, L., and C. Hall, *Family Fortunes: Men and Women of the English Middle Class, 1780–1850* (London: Routledge, 1987).

Davies, N. Z., *Society and Culture in Early Modern France: Eight Essays* (1965; Stanford, CA: Stanford University Press, 1975).

De la Beche, H., *A Geological Manual* (London, 1831).

De la Beche, H., and W. Conybeare, 'Notice of the Discovery of a New Fossil Animal, Forming a Link between the Ichthyosaurus and Crocodile, Together with General Remarks on the Osteology of the Ichthyosaurus', *Transactions of the Geological Society*, 5 (1821), pp. 559–94.

De Morgan, A., *A Budget of Paradoxes* (London: Longmans, Green & Co., 1872).

Dean, D., *Gideon Mantell and the Discovery of Dinosaurs* (Cambridge: Cambridge University Press, 1999).

Dear, P. (ed.), *The Literary Structure of Scientific Argument* (Philadelphia: University of Pennsylvania Press, 1991).

Desmond, A., and J. Moore, *Darwin* (New York: Warner Books, 1992).

Dettlebach, M., 'Humboldtian Science', in N. Jardine, J. A. Secord and E. Spary, *Cultures of Natural History* (Cambridge: Cambridge University Press, 1992), pp. 287–304.

Deutsch, H., and F. Nussbaum (eds), *'Defects': Engendering the Modern Body* (Ann Arbor, MI: University of Michigan Press, 2000).

Dibdin, T., *The Bibliomania; Or Book-Madness; Containing Some Account of the History, Symptoms, and Cure of This Fatal Disease; In an Epistle Addressed to Richard Heber, Esq.* (London, 1809).

—, *A Bibliographical, Antiquarian and Picturesque Tour in France and Germany*, 3 vols (London, 1821).

—, *Bibliographical, Antiquarian and Picturesque Tour in the Northern Counties of England and in Scotland*, 3 vols (London, 1838).

Dickens, C., *The Personal History of David Copperfield* (London: Chapman & Hall, 1850).

Dickinson, H. T., 'Spence, Thomas (1750–1814)', *Oxford Dictionary of National Biography* (Oxford: Oxford University Press, 2004).

Dircks, H., *Perpetuum mobile; Or, Search for Self-Motive Power, During the 17th, 18th, and 19th Centuries* (London: E. & F. N. Spon, 1861).

Dixon, H., *Saddle and Sirloin: Or, English Farm and Sporting Worthies* (London: Rogerson & Tuxford, 1870).

[Doran, J.], 'Things Not Generally Known: Curiosities of History; with New Lights. A Book for Old and Young. By John Timbs', *Athenaeum*, 1526 (1857), pp. 110–11.

—, 'The Romance of London: Strange Stories, Scenes, and Remarkable Persons of the Great Town. By John Timbs', *Athenaeum*, 1966 (1865), pp. 13–14.

Douglas, M., *Purity and Danger: An Analysis of the Concepts of Pollution and Taboo* (1966; London: Routledge, 2002).

Dryden, J., *The Conquest of Granada by the Spaniards* (London: Printed by T. N. for Henry Herringnan, 1672).

Duncan, J., and D. Gregory (eds), *Writes of Passage: Reading Travel Writing* (London and New York: Routledge, 1999).

Duncan, R., 'The Literary Gazette', in A. Sullivan (ed.), *British Literary Magazines: The Romantic Age, 1789–1836* (Westport, CT, and London: Greenwood Press, 1983), pp. 242–6.

Durrell, G., 'Foreword', in J. Blackburn, *Charles Waterton: 1782–1865* (London: Vintage, 1997), p. ix. Foreword dated 1988.

Dutton, D., and G. Nown, *Oddballs! Astonishing Tales of the Great Eccentrics* (London: Arrow Books, 1984).

Edgeworth, M., *Belinda*, 3 vols (London: J. Thomson, 1801).

—, *Patronage*, 4 vols (London: J. Johnson & Co., 1814).

Edginton, B., *Charles Waterton: A Biography* (Cambridge: The Lutterworth Press, 1996).

Edmonds, J. M., 'The Fossil Collection of the Misses Philpot of Lyme Regis', *Proceedings of the Dorset Natural History and Archaeological Society*, 98 (1976), pp. 43–8.

Eisenstein, E., *The Printing Press as an Agent of Change: Communications and Cultural Transformations in Early-Modern Europe* (Cambridge: Cambridge University Press, 1979).

Elfenbein, A., *Romantic Genius: The Prehistory of a Homosexual Role* (New York: Columbia University Press, 1999).

Elsner, J., and R. Cardinal, *The Cultures of Collecting* (London: Reaktion, 1994).

F.C., *Martinian Philosophy* (1828), broadside in *Billy Martin Philosopher's Nonsense*.

Fair Play, 'Letter to the Editor of the Tyne Mercury', *Tyne Mercury* (3 April 1827), p. 4.

—, 'Letter to the Editor of the Tyne Mercury', *Tyne Mercury* (10 April 1827), p. 3.

Fairholt, F., *Remarkable and Eccentric Characters, with Numerous Illustrations* (London: Richard Bentley, 1849).

Fara, P., 'Faces of Genius: Images of Isaac Newton in Eighteenth-Century England', in G. Cubitt and A. Warren (eds), *Heroic Reputations and Exemplary Lives* (Manchester: Manchester University Press, 2000), pp. 57–81.

Farber, P., 'The Development of Taxidermy and the History of Ornithology', *Isis*, 68 (1977), pp. 550–66.

—, *The Emergence of Ornithology as a Scientific Discipline, 1760–1850* (Dordrecht: Reidel, 1982).

Feaver, W., *The Art of John Martin* (Oxford: Clarendon Press, 1975).

Fineman, J., 'The History of the Anecdote: Fiction and Fiction', in H. Veser (ed.), *The New Historicism* (New York: Routledge, 1989), pp. 49–76.

Finley, G., 'The Deluge Pictures: Reflections on Goethe, J. M. W. Turner and Early Nineteenth-Century Science', *Zeitschrift für Kunstgeschichte*, 60 (1997), pp. 530–48.

Fish, S., *Is There a Text in This Class?: The Authority of Interpretive Communities* (Cambridge, MA: Harvard University Press, 1980).

Fitzgerald, P., *Memoirs of an Author* (London: Richard Bentley & Son, 1894).

Forrest, R., The Plesiosaur Site: www.plesiosaur.com (accessed 2004).

Forster, E., 'Just a Moment', *Evening Chronicle* (23 November 1972), p. 18.

Foucault, M., 'What Is an Author?', in D. Bouchard (ed.), *Language, Counter-Memory, Practice: Selected Essays and Interviews by Michel Foucault* (Oxford: Basil Blackwell, 1977), pp. 113–38.

—, *Discipline and Punish: The Birth of the Prison*, trans. A. Sheridan (first published in French 1975; Harmondsworth: Penguin, 1979).

Franklin, J., *The English Country House and Its Plan, 1835–1914* (London and Boston, MA: Rotuledge & Keegan Paul, 1981).

Frasca-Spada, M., 'The Many Lives of Eighteenth-Century Philosophy', *British Journal for the History of Philosophy*, 9 (2001), pp. 135–44.

Frasca-Spada, M., and N. Jardine (eds), *Books and the Sciences in History* (Cambridge: Cambridge University Press, 2000).

Friedman, A., and C. Donley, *Einstein as Myth and Muse* (Cambridge: Cambridge University Press, 1985).

Friedman, S., S. Dunwoody and C. Rogers (eds), *Scientists and Journalists: Reporting Science as News* (New York: The Free Press, 1986).

Frye, N., *Anatomy of Criticism: Four Essays* (Princeton, NJ: Princeton University Press, 1957).

Fyfe, A., 'Reading Natural History at the British Museum and the *Pictorial Museum*', in A. Fyfe and B. Lightman (eds), *Science in the Marketplace: Nineteenth-Century Sites and Experiences* (Chicago, IL: University of Chicago Press, 2007), pp. 196–230.

Fyfe, A., and B. Lightman (eds), *Science in the Marketplace: Nineteenth-Century Sites and Experiences* (Chicago, IL: University of Chicago Press, 2007).

G.B.R., 'Billy Martin', *Newcastle Courant* (31 January 1879), p. 6.

Gadamer, H.-G., *Truth and Method*, trans. J. Weinsheimer and D. G. Marshall (1960; London: Sheed & Ward, 1989).

Genette, G., *Narrative Discourse*, trans. J. E. Lewin (Oxford: Blackwell, 1980).

—, *Paratexts: Thresholds of Interpretation*, trans. J. E. Lewin (first published in French 1987; Cambridge: Cambridge University Press, 1997).

Genuth, S. S., *Comets, Popular Culture, and the Birth of Modern Cosmology* (Princeton, NJ: Princeton University Press, 1997).

Gerard, J., *Country House Life: Family and Servants, 1815–1914* (Oxford and Cambridge, MA: Blackwell, 1994).

Gibbons, R. compiler, *In Their Own Words: Autobiographical Writings of Seventeen of History's Greatest Thinkers and National Leaders* (New York: Random House, 1995).

Gill, M., 'Eccentricity and the Cultural Imagination in Nineteenth-Century France' (PhD dissertation, University of Oxford, 2004).

Gill, S., *Wordsworth and the Victorians* (Oxford: Clarendon Press, 1998).

Gilpin, W., *Three Essays: On Picturesque Beauty; on Picturesque Travel; and on Sketching Landscape: To Which Is Added a Poem, on Landscape Painting* (London: R. Bladmire, 1792).

Girouard, M., *Life in the English Country House: A Social and Architectural History* (New Haven, CT, and London: Yale University Press, 1978).

Gluckman, M., 'Rituals of Rebellion in South-East Africa', in Gluckman (ed.), *Order and Rebellion in Tribal Africa* (London: Cohen & West, 1963), pp. 110–36.

Goffman, E., *The Presentation of Self in Everyday Life* (1959; London: Penguin, 1990).

Gordon, *The Life and Correspondence of William Buckland* (London: John Murray, 1894).

Gossman, L., 'Anecdote and History', *History and Theory*, 42 (2003), pp. 143–68.

Granger, J., *A Biographical History of England, from Egbert the Great to the Revolution: Consisting of Characters Disposed in Different Classes, and Adapted to a Methodical Catalogue of Engraved British Heads. Intended as an Essay Towards Reducing Our Biography to a System, and a Help to the Knowledge of Portraits*, 3 vols (London: T. Davies, 1769).

Grasseni, C., 'Taxidermy as Rhetoric of Self-Making: Charles Waterton (1782–1865), Wandering Naturalist', *Studies in History and Philosophy of Biology and Biomedical Sciences*, 29 (1998), pp. 269–94.

Greenhalgh, P., *Ephemeral Vistas: The Expositions Universelles, Great Exhibitions and World's Fairs, 1851–1939* (Manchester: Manchester University Press, 1988).

Gregory, J., '"Local Characters": Eccentricity and the North-East in the Nineteenth Century', *Northern History*, 42 (2005), pp. 163–86.

—, 'Eccentric Lives: Character, Characters and Curiosities in Britain, *c.* 1760–1900', in W. Ernst (ed.), *Histories of the Normal and Abnormal: Social and Cultural Histories of Norms and Normativity* (London and New York: Routledge, 2006), pp. 73–100.

—, '"Local Characters" and Local, Regional and National Identities in Nineteenth-Century England and Scotland', in A. Brown (ed.), *Historical Perspectives on Social Identities* (Cambridge: Cambridge Scholars Press, 2006), pp. 45–60.

Gross, A., *The Rhetoric of Science* (Cambridge, MA: Harvard University Press, 1990).

Hall, S., 'Encoding/Decoding', in S. Hall, D. Hobson, A. Lowe and P. Willis (eds), *Culture, Media Language* (London: Hutchinson, 1980), pp. 128–62.

Hall, S. C., *Retrospect of a Long Life: From 1815 to 1883*, 2 vols (London: Richard Bentley & Son, 1883).

Hamilton, N., *Biography: A Brief History* (Cambridge, MA, and London: Harvard University Press, 2007).

Harley, G., 'Reminiscences of Charles Waterton', *Selbourne Magazine*, 2 (1889), pp. 20–3, 35–7, 114–17, 165–8.

Harrison, J. F. C., *The Second Coming: Popular Millenarianism 1780–1850* (New Brunswick, NJ: Rutgers University Press, 1979).

Hawkins, T., *Memoirs of Ichthyosauri and Plesiosauri, Extinct Monsters of the Ancient Earth, with Twenty-Eight Plates Copied from Specimens in the Author's Collection of Fossil Organic Remains* (London: Relfe & Fletcher, 1834).

—, *Memoirs of Ichthyosauri and Plesiosauri, Extinct Monsters of the Ancient Earth, with Twenty-Eight Plates, Copied from Specimens in the Author's Collection of Fossil Organic Remains*, 2nd edn (London: Relfe and Fletcher, 1835).

—, *The Book of the Great Sea-Dragons, Ichthyosauri and Plesiosauri,* גדלים חביבם*, Gedolim Taninim, of Moses; Extinct Monsters of the Ancient Earth; With Thirty Plates Copied from Skeletons in the Author's Collection of Fossil Organic Remains (Deposited in the British Museum)* (London: William Pickering, 1840).

—, *The Lost Angel and the History of the Old Adamites, Found Written on the Pillars of Seth: A Poem* (London: William Pickering, 1840).

—, *The Wars of Jehovah in Heaven, Earth, and Hell: In Nine Books* (London, 1844).

—, *My Life and Works Etc.* (London: The Chiswick Press, 1887).

Haynes, R., *From Faust to Strangelove: Representations of the Scientist in Western Literature* (Baltimore, MD, and London: Johns Hopkins University Press, 1994).

Hazlitt, W., *The Spirit of the Age: Or Contemporary Portraits* (1825; London: Oxford University Press, 1954).

[Hazlitt, W.], 'Of Persons One Would Wish to Have Seen', *New Monthly Magazine*, 16 (1826), pp. 32–41.

Head, G., *Forest Scenes and Incidents in the Wilds of North America: Being a Diary of a Winter's Route from Halifax to the Canadas, and During Four Months' Residence in the Woods on the Borders of Lakes Huron and Simcoe* (London: J. Murray, 1829).

—, *A Home Tour through the Manufacturing Districts of England, in the Summer of 1835* (1836; London and Edinburgh: Frank Cass & Co, 1968).

Henson, L., G. Cantor et al. (eds), *Culture and Science in the Nineteenth-Century Media* (Aldershot: Ashgate, 2004).

Heraud, J. A., 'The Wars of Jehovah, in Heaven, Earth and Hell', *Athenaeum*, 923 (1845), pp. 659–60.

Hewison, R., *The Heritage Industry: Britain in a Climate of Decline* (London: Methuen, 1987).

Higgins, D., *Romantic Genius and the Literary Magazine: Biography, Celebrity, Politics* (London: Routledge, 2005).

Hill, A., *The Fanciad. An Heroic Poem* (London: J. Osborn, 1743).

Hill, C., *The World Turned Upside Down: Radical Ideas During the English Revolution* (1972; London: Penguin Books, 1991).

Hilton, B., *A Mad, Bad and Dangerous People?: England 1783–1846* (Oxford: Clarendon Press, 2006).

Hobson, R., *Charles Waterton: His Home, Habits and Handiwork* (1866; London: Whittaker, 1867).

Holmes, R. 'The Proper Study?', in P. France and W. St Clair (eds), *Mapping Lives: The Uses of Biography* (Oxford: Oxford University Press, 2002), pp. 7–18.

Home, E., 'Some Account of the Fossil Remains of an Animal More Nearly Allied to Fishes Than Any of the Other Classes of Animals', *Philosophical Transactions of the Royal Society* (1814), pp. 571–7.

—, 'An Account of the Fossil Skeleton of the Proteo-Saurus', *Philosophical Transactions of the Royal Society* (1819), pp. 209–11.

—, 'Reasons for Giving the Name Proteo-Saurus to the Fossil Skeleton Which Has Been Described', *Philosophical Transactions of the Royal Society* (1819), pp. 212–16.

Houghton, W. (ed.), *The Wellesley Index to Victorian Periodicals* (Toronto: University of Toronto Press, 1979).

Howe, S., T. Sharpe and H. Torrens, *Ichthyosaurs: A History of Fossil 'Sea-Dragons'* (Cardiff: National Museum of Wales, 1981).

Hulbert, C., *The Select Museum of the World, or One Thousand Descriptions of Remarkable Antiquities, Curiosities, Beauties & Varieties of Nature & Art, in Asia, Africa, America & Europe* (Shrewsbury: C. Hulbert, 1822).

Hulbert, C., *Breakfast of Scraps* (Shrewsbury: C. Hulbert, date unknown).

Hulme, P., and Tim Youngs (eds), *The Cambridge Companion to Travel Writing* (Cambridge: Cambridge University Press, 2002).

Humboldt, A. von, *Personal Narrative of Travels to the Equinoctial Regions of the New Continent*, trans. H. M. Williams (London: Longman et al., 1814–29).

Humboldt, A von, and A. Bonpland, *Voyage aux régions équinoxiales du nouveau continent, fait dans les années 1799 à 1804*, 30 vols (Paris: Schoell, 1805–34).

Hunt, C. J., *The Book Trade in Northumberland and Durham to 1860* (Newcastle upon Tyne: Thorne's Students' Bookshop, 1975).

Hunt, C. J., and P. C. G. Isaac, 'The Regulation of the Booktrade in Newcastle Upon Tyne at the Beginning of the Nineteenth Century', *Archaeologia aeliana*, 5:5 (1977), pp. 163–78.

Hurley, K., *The Gothic Body: Sexuality, Materialism and Degeneration at the Fin De Siècle* (Cambridge: Cambridge University Press, 1996).

Hussey, C., *The Picturesque: Studies in a Point of View* (1927; London: Frank Cass & Co., 1967).

Hyams, E., *Capability Brown & Humphry Repton* (London: J. M. Dent & Sons Ltd., 1971).

Ingram, M., 'Ridings, Rough Music and Mocking Rhymes in Early Modern England', in B. Reay (ed.), *Popular Culture in Seventeenth-Century England* (London: Croom Helm, 1985), pp. 166–97.

Inkster, I., 'Science and the Mechanics' Institutes, 1820–1850: The Case of Sheffield', *Annals of Science*, 32 (1975), pp. 451–74.

Irwin, R. A., *Letters of Charles Waterton of Walton Hall, near Wakefield* (London: Rockliff Publishing Corporation, 1955).

J.H., 'William Martin', *The Northumbrian* (7 April 1883), p. 280.

Jacyna, L. S., 'Images of John Hunter in the Nineteenth Century', *History of Science*, 21 (1985), pp. 85–108.

Jardine, N., *The Scenes of Inquiry: On the Reality of Questions in the Sciences* (Oxford: Oxford University Press, 2000).

Jardine, N., J. A. Secord and E. Spary (eds), *Cultures of Natural History* (Cambridge: Cambridge University Press, 1992).

Jauss, H. R., *Towards an Aesthetics of Reception*, trans. T. Bahti (Minneapolis: University of Minnesota Press, 1982).

[Jeaffreson, J.], 'Stories of Inventors and Discoveries. By John Timbs', *Athenaeum*, 1675 (1859), p. 739.

—, 'English Eccentrics and Eccentricities. By John Timbs', *Athenaeum*, 2049 (1867), p. 155.

John, J., *Eccentrics* (London: Duckworth, 2001).

Johns, A., *The Nature of the Book: Print and Knowledge in the Making* (Chicago, IL: University of Chicago Press, 1998).

Joliffe, J., *Eccentrics* (London: Duckworth, 2001).

Kahn, M., *Narrative Transvestism: Rhetoric and Gender in the Eighteenth-Century English Novel* (Ithaca, NY, and London: Cornell University Press, 1991).

Keats, J., *Lamia* (1819; London: Kessinger Publishing, 2004).

Keay, J., *Eccentric Travellers* (1982; London: John Murray, 2001).

Kendall, J., *Eccentricity; or a Check to Censoriousness; with Chapters on Other Subjects* (London: Simpkin, Marshall & Co., 1859).

[Kinsey, W.], 'Random Recollections of a Visit to Walton Hall', *Gentleman's Magazine*, 29 (1848), pp. 33–9.

Kirby, R. S., *Kirby's Wonderful and Eccentric Museum; Or, Magazine of Remarkable Characters. Including All the Curiosities of Nature and Art, from the Remotest Period to the Present Time, Drawn from Every Authentic Source. Illustrated with One Hundred and Twenty-Four Engravings. Chiefly Taken from Rare and Curious Prints or Original Drawings*, 6 vols (London: R. S. Kirby, 1803–20).

Klancher, J., 'British Theory and Criticism: Romantic Period and Early Nineteenth Century', in M. Groden and M. Kreiswirth (eds), *The Johns Hopkins Guide to Literary Theory and Criticism* (Baltimore, MD: Johns Hopkins University Press, 1997), pp. 112–15.

Klein, L. E., 'Gender and the Public/Private Distinction in the Eighteenth Century: Some Questions About Evidence and Analytic Procedure', *Eighteenth-Century Studies*, 29 (1995), pp. 97–109.

Knight, C., *Knight's Excursion-Train Companion* (London: Charles Knight, 1851).

Knox, K., 'Lunatick Visions: Prophecy, Signs and Scientific Knowledge in 1790s London', *History of Science*, 37 (1999), pp. 427–58.

Korte, B., *English Travel Writing from Pilgrimages to Postcolonial Explorations*, trans. C. Matthias (Basingstoke: Macmillan Press Ltd, 2000).

Langford, P., *Englishness Identified: Manners and Character 1650–1850* (Oxford: Oxford University Press, 2000).

Langhorne, J. B., 'W. Martin, the Natural Philosopher', *Notes and Queries*, 4:12 (1873), p. 134.

Latimer, J., *Local Records; Or, Historical Register of Remarkable Events, Which Have Occurred in Northumberland & Durham, Newcastle-Upon-Tyne, and Berwick-Upon-Tweed* ... (Newcastle: The Chronicle Office, 1857).

Leask, N., *Curiosity and the Aesthetics of Travel Writing 1770–1840* (2002; Oxford: Oxford University Press, 2004).

Lee, H., *Body Parts: Essays in Life-Writing* (London: Chatto & Windus, 2005).

Lee, S., *Taxidermy: Or, the Art of Collecting, Preparing, and Mounting Objects of Natural History. For the Use of Museums and Travellers* (1821; London: Longman et al., 1843).

Lever, C., 'Father of Wildlife Conservation', *Country Life*, 171 (1982), pp. 1698–9.

Lindberg, D., and R. Numbers (eds), *God and Nature: Historical Essays on the Encounter between Christianity and Science* (Berkeley: University of California Press, 1986).

Lonsdale, B., 'A Fantastic Squire at Walton Hall', *The Dalesman*, 27 (1965), pp. 120–2.

[Lord, P. B.], 'Memoirs of Ichthyosauri and Plesiosauri Etc.', *Athenaeum*, 347 (1834), pp. 469–70.

Lowenthal, D., *The Past Is a Foreign Country* (Cambridge: Cambridge University Press, 1985).

Luckhurst, M., and J. Moody (eds), *Theatre and Celebrity in Britain, 1660–2000* (London: Palgrave Macmillan, 2005).

Lyell, C., *Principles of Geology* (1830–3; London: Penguin, 1997).

McCalman, I., *Radical Underworld: Prophets, Revolutionaries and Pornographers in London, 1795–1840* (Cambridge: Cambridge University Press, 1988).

—, 'New Jerusalems: Prophecy, Dissent and Radical Culture in England, 1786–1830', in K. Haakonssen (ed.), *Enlightenment and Religion: Rational Dissent in Eighteenth-Century Britain* (Cambridge: Cambridge University Press, 1997), pp. 312–35.

McGowan, C., *The Dragon Seekers* (Cambridge, MA: Perseus Publishing, 2001).

Mackay, C., *Memoirs of Extraordinary Popular Delusions*, 3 vols (London: R. Bentley, 1841).

McKendrick, N., 'George Packwood and the Commercialization of Shaving: The Art of Eighteenth-Century Advertising or "the Way to Get Money and Be Happy"', in N. McKendrick, J. Brewer and J. H. Plumb (eds), *The Birth of a Consumer Society: The Commercialization of Eighteenth-Century England* (Bloomington, IN: Indiana University Press, 1982), pp. 145–94.

McKenzie, D. F., *Bibliography and the Sociology of Texts* (London: British Library, c. 1986).

Mackenzie, E., *A Descriptive and Historical Account of the Town and County of Newcastle Upon Tyne: Including the Borough of Gateshead* (Newcastle upon Tyne: Mackenzie and Dent, 1827).

MacKenzie, J., 'The Iconography of the Exemplary Life: The Case of David Livingstone', in G. Cubitt and A. Warren (eds), *Heroic Reputations and Exemplary Lives* (Manchester: Manchester University Press, 2000), pp. 84–104.

Mandler, P., *The Fall and Rise of the Stately Home* (New Haven, CT, and London: Yale University Press, 1997).

Mantell, G., 'Notice on the Iguanodon, a Newly Discovered Fossil Reptile, from the Sandstone of Tilgate Forest, in Sussex', *Philosophical Transactions of the Royal Society of London*, 115 (1825), pp. 179–86.

—, *Wonders of Geology* (London, 1838).

Martin, J., *The Life of Jonathan Martin, of Darlington, Tanner: Written by Himself, etc.* (Darlington: Thomas Thompson, 1825).

Martin, P. J., *A Geological Memoir on Part of Western Sussex; with Some Observations Upon Chalk-Basins, the Weald-Denudation, and Outliers-by-Protrusion* (London, 1828).

Martin, W., *A New System of Natural Philosophy, on the Principle of Perpetual Motion; with a Variety of Other Useful Discoveries. Patronised by His Grace the Duke of Northumberland* (Newcastle: Preston & Heaton, 1821).

—, *Belshazzar's Feast*, broadside dated 26 December 1826 (Newcastle: Edward Walker), in *Billy Martin Philosopher's Nonsense*.

—, *Lines on the Safety Lamp; With a Reply to an Attack Headed 'Ecce Homo', Written by a Fellow Who Calls Himself 'A Newtonian'* (1827), broadside in *Billy Martin Philosopher's Nonsense*.

—, *The Downfall of the Newtonian False System of Philosophy* (1827), broadside in *Billy Martin Philosopher's Nonsense*.

—, *A New Philosophical Song or Poem Book, Called the Northumberland Bard; or, the Downfall of All False Philosophy* (Newcastle: Thomas Blagburn, 1827).

—, *The Downfall of the Newtonian System* (Newcastle: Edgar, 1827), in *Billy Martin Philosopher's Nonsense*.

—, *Martin, Natural Philosopher!* (Newcastle: Edward Walker, 1827), broadside in *Billy Martin Philosopher's Nonsense*.

—, 'New System of Philosophy', *Newcastle Courant* (19 January 1828), p. 1.

—, 'New System of Philosophy', *Newcastle Courant* (15 March 1828), p. 1.

—, 'New System of Philosophy', *Newcastle Courant* (29 March 1828), p. 1.

—, 'New System of Philosophy', *Newcastle Courant* (5 April 1828), p. 1.

—, 'New System of Philosophy', *Newcastle Courant* (19 April 1828), p. 1.

—, 'New System of Philosophy', *Newcastle Courant* (10 May 1828), p. 1.

—, 'New System of Philosophy', *Newcastle Courant* (17 May 1828), p. 1.

—, 'New System of Philosophy', *Newcastle Courant* (24 May 1828), p. 1.

—, 'New System of Philosophy', *Newcastle Courant* (31 May 1828), p. 1.

—, 'New System of Philosophy', *Newcastle Courant* (7 June 1828), p. 1.

—, 'New System of Philosophy', *Newcastle Courant* (14 June 1828), p. 1.

—, 'New System of Philosophy', *Newcastle Courant* (21 June 1828), p. 1.

—, *A Challenge to the Whole World to Produce a Travelling Machine Equal to this Invention*, broadside dated 26 August 1828 (Newcastle: Edward Walker, 1828), in *Billy Martin Philosopher's Nonsense.*

—, *William Martin's Challenge to the Whole Terrestrial Globe, as a Philosopher and Critic, and Poet and Prophet; Shewing the Travels of His Mind, the Quick Motion of the Soul, That Never-Dying Principle, the Spirit Belonging to Mortal Man* (Newcastle upon Tyne: For the author by Wm Fordyce, 1829).

—, *The Defeat of Learned Humbugs, and the Downfall of All False Philosophers, in the Nineteenth Century, for the Good of All Mankind, and the Christian Church* (Newcastle upon Tyne, 1832).

—, *A Short Outline of the Philosopher's Life, from Being a Child in Frocks to This Present Day, after the Defeat of All Impostors, False Philosophers, since the Creation; by the Will of the Mighty God of the Universe, He Has Laid the Grand Foundation for Church Reform by True Philosophy &C.* (Newcastle: J. Blackwell and Co., 1833).

—, *The Christian Philosopher's Explanation of the General Deluge, and the Proper Cause of All the Different Strata: Wherein It Is Clearly Demonstrated, That One Deluge Was the Cause of the Whole, Which Divinely Proves That God Is Not a Liar, but That the Bible Is Strictly True* (Newcastle: W. Fordyce, 1834).

—, *The Thunder Storm of Dreadful Forked Lightning: God's Judgment against All False Teachers, That Cause the People to Err, and Those That Are Led by Them Are Destroyed, According to God's Word. Including an Account of the Railway Phenomenon, the Wonder of the World!* (Newcastle upon Tyne: Pattison & Ross, 1837).

—, *The Total Defeat of the Yearly Meeting of the British Association of Scientific Asses* (Newcastle: Pattison & Ross, 1838).

—, *Prophetic Knowledge from That Spirit of God Which Directed Daniel the Prophet to Interpret the Hand Writing on the Wall to the Impious King Belshazzar* (Newcastle: Pattison & Ross, 1839).

May, R., *News from Wallsend* (Newcastle, 1827), broadside in *Billy Martin Philosopher's Nonsense.*

Menteath, J., 'Some Account of Walton Hall, the Seat of Charles Waterton, Esq', *Magazine of Natural History*, 8 (1834), pp. 28–36.

—, 'Walton Hall', *The Youth's Instructor and Guardian*, 19 (1835), pp. 92–7, 125–9.

Meredith, G., *The Ordeal of Richard Feverel: A History of Father and Son* (London: Chapman & Hall, 1859).

Metzner, P., *Crescendo of the Virtuoso: Spectacle, Skill, and Self-Promotion in Paris during the Age of Revolution* (Berkeley: University of California Press, 1998).

Michell, J., *Eccentric Lives and Peculiar Notions* (London: Thames & Hudson, 1989).

Mighall, R., *A Geography of Victorian Gothic Fiction: Mapping History's Nightmares* (Oxford: Oxford University Press, 1999).

Mill, J. S., *On Liberty* (1859; London: David Campbell, 1992).

[Mill, J. S.], 'The Spirit of the Age', in A. Robson (ed.), *Collected Works of John Stuart Mill, Vol. 22: Newspaper Writings by John Stuart Mill, December 1822–July 1831* (London: Routledge, 1831), pp. 227–316.

Miller, D., 'True Myths: James Watt's Kettle, His Condenser, and His Chemistry', *History of Science*, 42 (2004), pp. 333–60.

Miller, D., and P. Reill (eds), *Visions of Empire: Voyages, Botany and Representations of Nature* (Cambridge: Cambridge University Press, 1996).

Miller, D., J. Kitzinger and P. Beharrell, *The Circuit of Mass Communication: Media Strategies, Representation and Audience Reception in the Aids Crisis* (London: Sage Publications, 1998).

Miller, H., *The Old Red Sandstone; or New Walks in an Old Field* (Edinburgh: John Johnstone, 1841).

Milne, M., 'Periodical Publishing in the Provinces: The Mitchell Family of Newcastle Upon Tyne', *Victorian Periodicals Newsletter*, 10 (1977), pp. 174–82.

[Mitford, J.], 'The Lost Angel, and the History of the Old Adamites, Found Written on the Pillars of Seth', *Gentleman's Magazine*, 15 (1841), pp. 169–72.

—, 'The Wars of Jehovah in Heaven, Earth and Hell', *Gentleman's Magazine*, 23 (1845), pp. 516–19.

Moir, E., *The Discovery of Britain: The English Tourists 1540–1840* (London: Routledge & Keegan Paul, 1964).

Moore, J., 'Geologists and Interpreters of Genesis in the Nineteenth Century', in D. Lindberg and R. Numbers (eds), *God and Nature: Historical Essays on the Encounter between Christianity and Science* (Berkeley: University of California Press, 1986), pp. 322–50.

Morley, D., *Television, Audiences and Cultural Studies* (London: Routledge, 1992).

Morrell, J., 'Professionalisation', in R. C. Olby et al. (eds), *Companion to the History of Modern Science* (London and New York: Routledge, 1990), pp. 980–9.

Myers, G., *Writing Biology: Texts in the Social Construction of Scientific Knowledge* (Madison, WI: University of Wisconsin Press, 1990).

Nelkin, D., *Selling Science: How the Press Covers Science and Technology* (New York: W. H. Freeman & Co., 1987).

Nettleship, J., 'A Yorkshire Naturalist's Debt to Lancashire', *Yorkshire Ridings Magazine*, 19 (1982), p. 77.

Newtonian, *Ecce Homo* (Newcastle: Blagburn, 1827), broadside in *Billy Martin Philosopher's Nonsense*.

Nicholas, M., *The World's Greatest Cranks & Crackpots* (London: Octopus Books Limited, 1984).

Nord, D. E., '"Marks of Race": Gypsy Figures and Eccentric Femininity in Nineteenth-Century Women's Writing', *Victorian Studies*, 41 (1988), pp. 191–210.

O'Connor, R., 'The Poetics of Geology: A Science and Its Literature in Britain, 1802–1856' (PhD dissertation: University of Cambridge, 2003).

—, 'Thomas Hawkins and Geological Spectacle', *Proceedings of the Geologists' Association*, 114 (2003), pp. 227–41.

—, *The Earth on Show: Fossils and the Poetics of Popular Science, 1802–1856* (Chicago, IL: Chicago University Press, 2008).

Oppenheim, J., *The Other World: Spiritualism and Psychical Research in England, 1850–1914* (Cambridge: Cambridge University Press, 1985).

Orange, D., 'Rational Dissent and Provincial Science: William Turner and the Newcastle Literary and Philosophical Society', in I. Inkster and J. Morrell (eds), *Metropolis and Province: Science in British Culture, 1780–1850* (London: Hutchinson, 1983), pp. 205–30.

Ousby, I., *The Englishman's England: Taste, Travel and the Rise of Tourism* (Cambridge: Cambridge University Press, 1990).

Owen, R., *The Life of Richard Owen*, 2 vols (London: John Murray, 1894).

Paley, M., *The Apocalyptic Sublime* (New Haven, CT: Yale University Press, 1986).

—, *Apocalypse and Millennium in English Romantic Poetry* (Oxford: Clarendon Press, 1999).

Paradis, J., 'The Natural Historian as Antiquary of the World: Hugh Miller and the Rise of Literary Natural History', in M. Shortland (ed.), *Hugh Miller and the Controversies of Victorian Science* (Oxford: Clarendon Press, 1996), pp. 122–50.

Parrinder, P., *Authors and Authority: A Study of English Literary Criticism and Its Relation to Culture 1750–1900* (London: Routledge & Keegan Paul, 1977).

Pateman, C., *The Disorder of Women: Democracy, Feminism and Political Theory* (Cambridge: Polity Press, 1989).

Pearce, S., *On Collecting: An Investigation into Collecting in the European Tradition* (London and New York: Routledge, 1995).

Penn, G., *Comparative Estimate of the Mineral and Mosaical Geologies* (London, 1822).

Pendered, M., *John Martin, Painter: His Life and Times* (London: Hurst & Blackett, 1923).

Perkins, M., *Visions of the Future: Almanacs, Time, and Cultural Change 1775–1870* (Oxford: Clarendon Press, 1996).

Phillips, J., *A Guide to Geology* (London: Longman et al., 1834).

Pickover, C., *Strange Brains and Genius: The Secret Lives of Eccentric Scientists and Madmen* (New York: Quill, 1999).

Pointon, M., *Hanging the Head: Portraiture and Social Formation in Eighteenth-Century England* (New Haven, CT, and London: Yale University Press, 1993).

Pomian, K., *Collectors and Curiosities: Paris and Venice, 1500–1800*, trans. E. Wiles-Porter (first published in French 1987; Cambridge: Polity Press, 1990).

Porter, R., 'Gentlemen and Geology: The Emergence of a Scientific Career, 1660–1920', *The Historical Journal*, 21: 4 (1978), pp. 809–36.

Pratt, M. L., *Imperial Eyes: Studies in Travel Writing and Transculturation* (London and New York: Routledge, 1992).

Price, L., *The Anthology and the Rise of the Novel: From Richardson to George Eliot* (Cambridge: Cambridge University Press, 2000).

Purcell, R. W., and S. J. Gould, *Finders, Keepers: Eight Collectors* (London: Hutchinson Radius, 1992).

Qureshi, S., 'Living Curiosities: Human Ethnological Display, 1800–1855' (PhD dissertation: University of Cambridge, 2005).

[Raspe, R. E.], *Surprising Adventures of the Renowned Baron Munchausen, Containing Singular Travels, Campaigns, Voyages, and Adventures. Also, an Account of a Voyage to the Moon and Dog Star* (1785; London: Thomas Tegg, 1811).

Raven, J., H. Small and N. Tadmor, 'Introduction', in Raven, Small and Tadmor (eds), *The Practice and Representation of Reading in England* (Cambridge: Cambridge University Press, 1996), pp. 1–21.

Reach, A., *The Natural History of Bores* (London: D. Bogue, 1847).

Reader's Digest, *Great British Eccentrics: They Entertained, Exasperated and Charmed a Nation* (London: The Reader's Digest Association, 1982).

Recorde, R., *The Castle of Knowledge, Containing the Explication of the Sphere Both Celestiall and Materiall, Etc.* (London, 1556).

Reid, F., 'Isaac Frost's *Two Systems of Astronomy* (1846): Plebeian Resistance and Scriptural Astronomy', *British Journal for the History of Science*, 38 (2005), pp. 161–77.

Riffaterre, M., *Text Production*, trans. T. Lyons (first published in French in 1979; New York: Columbia University Press, 1983).

Ritchie, C., 'A Prophet at Walton Hall', *The Dalesman*, 44 (1983), pp. 763–5.

Ritvo, H., *The Platypus and the Mermaid and Other Figments of the Classifying Imagination* (Cambridge, MA: Harvard University Press, 1997).

[Robinson, C.], 'Letter to Christopher North, Esquire, on the Spirit of the Age', *Blackwood's Edinburgh Magazine*, 28 (1830), p. 900.

Roslynn, H., *From Faust to Strangelove: Representations of the Scientist in Western Literature* (Baltimore, MD: Johns Hopkins University Press, 1994).

Rowland, W., *Literature and the Marketplace: Romantic Writers and Their Audiences in Great Britain and the United States* (Lincoln, NE, and London: University of Nebraska Press, 1996).

Rudwick, M., 'The Emergence of a Visual Language for Geological Science, 1760–1840', *History of Science*, 14 (1976), pp. 149–95.

—, 'The Shape and Meaning of Earth History', in D. Lindberg and R. Numbers (eds), *God and Nature: Historical Essays on the Encounter between Christianity and Science* (Berkeley: University of California Press, 1986), pp. 296–321.

—, *Scenes from Deep Time: Early Pictorial Representations of the Prehistoric World* (Chicago, IL, and London: The University of Chicago Press, 1992).

Rupke, N., *The Great Chain of History: William Buckland and the English School of Geology (1814–1849)* (Oxford: Clarendon Press, 1983).

Russell, W., *Eccentric Personages*, 2 vols (London: John Maxwell & Co., 1864).

St Clair, W., *The Reading Nation in the Romantic Period* (Cambridge: Cambridge University Press, 2004).

Sambrook, J., *English Pastoral Poetry* (Boston, MA: Twayne Publishers, 1983).

Samuel, R., *Theatres of Memory. Volume 1: Past and Present in Contemporary Culture* (London: Verso, 1994).

Savage, R., 'The Bastard', in S. Johnson (ed.), *The Works of Richard Savage* (London: T. Evans, 1777), pp. 91–5.

Shapin, S., *A Social History of Truth: Civility and Science in Seventeenth-Century England* (Chicago, IL: University of Chicago Press, 1994).

Schachterle, L., 'The Metropolitan Magazine', in A. Sullivan (ed.), *British Literary Magazines: The Romantic Age 1789–1836* (Westport, CT, and London: Greenwood Press, 1983), pp. 304–8.

Schaffer, S., 'Newton's Comets and the Transformation of Astrology', in P. Curry (ed.), *Astrology, Science and Society: Historical Essays* (Woodbridge: The Boydell Press, 1987), pp. 219–43.

—, 'Genius in Romantic Natural Philosophy', in N. Jardine and A. Cunningham (eds), *Romanticism and the Sciences* (Cambridge: Cambridge University Press, 1990), pp. 82–98.

Schechner Genuth, S., *Comets, Popular Culture, and the Birth of Modern Cosmology* (Princeton, NJ: Princeton University Press, 1997).

Schivelbusch, W., *The Railway Journey: Trains and Travel in the 19th Century* (Oxford: Basil Blackwell, 1977).

Scott, W. B., *Autobiographical Notes of the Life of William Bell Scott*, ed. W. Minto, 2 vols (London: James R. Osgood, McIlbaine & Co., 1892).

Secord, A., 'Corresponding Interests: Artisans and Gentlemen in Nineteenth-Century Natural History', *British Journal of the History of Science*, 27 (1994), pp. 383–408.

—, 'Science in the Pub: Artisan Botanists in Early Nineteenth-Century Lancashire', *History of Science*, 32 (1994), pp. 269–315.

Secord, J. A., *Victorian Sensation: The Extraordinary Publication, Reception, and Secret Authorship of Vestiges of the Natural History of Creation* (Chicago, IL, and London: University of Chicago Press, 2000).

—, 'Scrapbook Science: Composite Caricures in Late Georgian England', in A. Shteir and B. Lightman (eds), *Figuring It Out: Science, Gender, and Visual Culture* (Hanover, NH: Dartmouth College Press, 2006), pp. 164–91.

Seed, D., 'Nineteenth-Century Travel Writing: An Introduction', *The Yearbook of English Studies (Nineteenth-Century Travel Writing)*, 34 (2004), pp. 1–5.

Shapin, S., and B. Barnes, 'Science, Nature and Control: Interpreting Mechanics' Institutes', *Social Studies of Science*, 7 (1977), pp. 31–74.

Shaw, K., *The Mammoth Book of Oddballs and Eccentrics* (London: Robinson, 2000).

Shesgreen, S., *Images of the Outcast: The Urban Poor in the Cries of London* (Manchester: Manchester University Press, 2002).

Shoemaker, R., *Gender in English Society, 1650–1850: The Emergence of Separate Spheres?* (London and New York: Longman, 1998).

Shteir, A., *Cultivating Women, Cultivating Science: Flora's Daughters and Botany in England, 1760 to 1860* (Baltimore, MD: Johns Hopkins University Press, 1996).

Siegel, J., *Desire and Excess: The Nineteenth-Century Culture of Art* (Princeton, NJ, and Oxford: Princeton University Press, 2000).

Silverstone, R., *Framing Science: The Making of a BBC Documentary* (London: BFI Publishing, 1985).

—, *Television and Everyday Life* (London: Routledge, 1994).

Sitwell, E., *The English Eccentrics* (London: Faber & Faber, 1933).

Smith, A., *The Natural History of Stuck-Up People* (London, 1847).

Smith, J. D., and T. Smith, *South and West Yorkshire Curiosities* (Wimborne: Dovecote Press, 1992).

Smith, J. T., *Vagabondiana; Or, Anecdotes of Mendicant Wanderers through the Streets of London; with Portraits of the Most Remarkable Drawn from Life* (London: Published for the proprietor, 1817).

—, *The Cries of London: Exhibiting Several of the Itinerant Traders of Ancient and Modern Times. Copied from Rare Engravings, or Drawn from the Life, by John Thomas Smith, Late Keeper of the Prints in the British Museum. With a Memoir and Portrait of the Author* (London: John Bowyer Nichols & Son, 1839).

[Smith, S.], 'Wanderings in South America Etc.', *Edinburgh Review*, 43 (1826), pp. 299–315.

Smiles, S., *Self Help: With Illustrations of Character and Conduct* (London, John Murray, 1859).

Smith, W. (ed.), *Old Yorkshire* (London: Longmans, Green, 1882).

Sommer, M., 'The Romantic Cave? The Scientific and Poetic Quests for Subterranean Spaces in Britain', *Earth Sciences History*, 22 (2003), pp. 172–208.

[Southern, H.], 'Waterton's Wanderings in South America', *London Magazine*, 4 (1826), pp. 343–53.

Spurgeon, D., 'The Athenaeum', in A. Sullivan (ed.), *British Literary Magazines: The Romantic Age 1789–1836* (Westport, CT, and London: Greenwood Press, 1983), pp. 21–4.

Stallybrass, P., and A. White, *The Politics and Poetics of Transgression* (London: Menthuen, 1986).

Star, S. L., 'Craft Vs. Commodity, Mess Vs. Transcendence: How the Right Tool Became the Wrong One in the Case of Taxidermy and Natural History', in A. Clarke and J. Fujimura (eds), *The Right Tools for the Job: At Work in Twentieth-Century Life Sciences* (Princeton, NJ: Princeton University Press, 1992), pp. 257–86.

[Stephens, F.], 'Anecdote Biography. By John Timbs', *Athenaeum*, 1724 (1860), p. 627.

Stewart, S., *On Longing: Narratives of the Miniature, the Gigantic, the Souvenir, the Collection* (Durham, NC, and London: Duke University Press, 1993).

Suleiman, S., and I. Crossman, *The Reader in the Text: Essays on Audience and Interpretation* (Princeton, NJ: Princeton University Press, 1980).

Sussman, H., *Victorian Masculinities: Manhood and Masculine Poetics in Early Victorian Literature and Art* (Cambridge: Cambridge University Press, 1995).

Sutcliffe, J., *A Short Introduction to the Study of Geology; Comprising a New Theory of the Elevation of the Mountains and the Stratification of the Earth: In Which the Mosaic Account of the Creation and the Deluge is Vindicated* (London, 1817–19).

Swainson, W., *On the Natural History and Classification of Birds* (London: Longman, 1836).

Sysak, J., 'Coleridge's Construction of Newton', *Annals of Science*, 50 (1993), pp. 59–81.

Taylor, F., 'Yorkshire's Naturalist', *The Dalesman*, 19 (1957), pp. 13–14.

Taylor, M., 'Thomas Hawkins FGS', *Geological Curator*, 5 (1987), pp. 112–14.

—, 'Thomas Hawkins of "The Great Sea Dragons": The Isle of Wight Connection', *Geological Society of the Isle of Wight Newsletter*, 1: 5 (1997), pp. 5–10.

—, 'Mary Anning, Thomas Hawkins and Hugh Miller, and the Realities of Being a Provincial Fossil Collector', *Edinburgh Geologist*, 34 (2000), pp. 28–37.

—, 'Joseph Clark III's Reminiscences About the Somerset Fossil Reptile Collector Thomas Hawkins (1810–1889)', *Somerset Archaeology and Natural History*, 146 (2002), pp. 1–10.

—, 'Thomas Hawkins', *Oxford Dictionary of National Biography* (Oxford: Oxford University Press, 2004).

—, *Hugh Miller: Stonemason, Geologist, Writer* (Edinburgh: National Museums Scotland, 2007).

Thackray, J., and B. Press, *The Natural History Museum: Nature's Treasurehouse* (London: Natural History Museum, 2001).

Thompson, E. P., *Customs in Common* (London: The Merlin Press, 1991).

Ticklecheek, T., *The Cries of London, Displaying the Manners, Customs & Characters, of Various People Who Traverse London Streets with Articles to Sell. To Which Is Added Some Pretty Poetry Applicable to Each Character. Intended to Amuse and Instruct All Good Children, with London and the Country Contrasted* (London: J. Fairburn, 1797).

Timbs, J., *English Eccentrics and Eccentricities*, 2 vols (London: Richard Bentley, 1866).

—, *Eccentricities of the Animal Creation* (London: Seeley, Jackson & Halliday, 1869).

—, 'My Autobiography', *The Leisure Hour* (1871).

Timpson, J., *Timpson's English Eccentrics* (Norwich: Jarrold Publishing, 1991).

Tinniswood, A., *A History of Country House Visiting: Five Centuries of Tourism and Taste* (Oxford: Basil Blackwell and the National Trust, 1989).

Todorov, T., *The Fantastic: A Structural Approach to a Literary Genre* (Ithaca, NY: Cornell University Press, 1975).

Tompkins, J. (ed.), *Reader-Response Criticism: From Formalism to Post-Structuralism* (Baltimore, MD: John Hopkins University Press, 1980).

Topham, J., 'Scientific Publshing and the Reading of Science in Nineteenth-Century Britain: A Historiographical Survey and Guide to Sources', *Studies in History and Philosophy of Science*, 31A (2000), pp. 559–612.

—, 'The *Mirror of Literature, Amusement and Instruction* and Cheap Miscellanies in Early Nineteenth-Century Britain', in G. Cantor et al. (eds), *Science in the Nineteenth-Century Periodical: Reading the Magazine of Nature* (Cambridge: Cambridge University Press, 2004), pp. 37–66.

Torrens, H., 'Mary Anning (1799–1847) of Lyme: "The Greatest Fossilist the World Ever Knew"', *British Journal for the History of Science*, 28 (1995), pp. 257–84.

Torrens, H., et al., 'Etheldred Benett of Wiltshire, England, the First Lady Geologist', *Proceedings of the Academy of Natural Sciences of Philadelphia*, 150 (2000), pp. 59–123.

Turner, V., *The Ritual Process: Structure and Anti-Structure* (London: Routledge & Keegan Paul, 1969).

—, *The Anthropology of Performance* (1987; New York: PAJ Publications, 1988).

Tweedie, E. B., *George Harley FRS or the Life of a London Physician* (London: The Scientific Press, 1899).

Uglow, J., *Elizabeth Gaskell: A Habit of Stories* (London: Faber & Faber, 1993).

—, *Nature's Engraver: A Life of Thomas Bewick* (London: Faber, 2006).

Vickery, A., 'Golden Age to Separate Spheres? A Review of the Categories and Chronology of English Women's History', *Historical Journal*, 36 (1993), pp. 383–414.

Vizetelly, H., *Glances Back through Seventy Years: Autobiographical and Other Reminiscences*, 2 vols (London: Kegan Paul, Trench, Trübner & Co., 1893).

W.N., 'Character of Robert Lecky Esq', *The Times* (26 June 1787), p. 1.

Wakefield Museum, *Charles Waterton 1782–1865: Traveller and Naturalist* (Wakefield: Wakefield Museum, 1982).

Wallace, A., *Walking, Literature, and English Culture: The Origins and Uses of Peripatetic in the Nineteenth Century* (Oxford: Clarendon Press, 1993).

Wallis, P. J., *The Book Trade in Northumberland and Durham to 1860: A Supplement to C. J. Hunt's Biographical Dictionary* (Newcastle upon Tyne: Thorne's Bookshops, 1981).

Waterton, C., *Wanderings in South America, the North-West of the United States and the Antilles, in the Years 1812, 1816, 1820 & 1824, with Original Instructions for the Perfect Preservation of Birds, Etc. for Cabinets of Natural History* (London: Mawman, 1825).

Waterton, C., *Essays on Natural History, Chiefly Ornithology* (London: Longman, 1838).

—, *Essays on Natural History, Chiefly Ornithology* (2nd series) (London: Longman, 1844).

—, *Essays on Natural History, Chiefly Ornithology* (3rd series) (London: Longman, 1857).

—, *Essays on Natural History, Chiefly Ornithology. Edited, with a Life of the Author by N. Moore* (London: Frederick Warne & Co., 1871).

Watkin, D., *The English Vision: The Picturesque in Architecture, Landscape and Garden Design* (London: John Murray, 1982).

—, 'Monuments and Mausolea in the Age of Enlightenment', in G. Waterfield (ed.), *Soane and Death: The Tombs and Monuments of Sir John Soane* (London: Dulwich Picture Gallery, 1996), pp. 9–25.

Weaver, T., 'Memoir on the Geological Relations of the East of Ireland', *Transactions of the Geological Society*, 5 (1821), pp. 117–304.

Webster, C., *The Great Instauration: Science, Medicine and Reform 1626–1660* (London: Duckworth, 1975).

[Weddall, W.], 'Memoirs of Ichthyosauri and Plesiosauri', *Gentleman's Magazine*, 2 (1834), pp. 121–4.

Weeks, D., and J. James, *Eccentrics* (London: Weidenfeld & Nicolson, 1995).

Welford, R., *Men of Mark 'Twixt Tyne and Tweed* (London: Walter Scott, 1895).

West, J., 'The Squire Brought 'Em Back Alive', *Yorkshire Life Illustrated* (September 1958), pp. 22–3.

Whye, J. van, *Phrenology and the Origins of Victorian Scientific Naturalism* (Aldershot: Ashgate, 2004).

Wiesenthal, C., *Figuring Madness in Nineteenth-Century Fiction* (Basingstoke: Macmillan, 1997).

Williams, R., *Keywords: A Vocabulary of Culture and Society* (London: Croom Helm, 1976).

Wilson, G. H., *The Eccentric Mirror: Reflecting a Faithful and Interesting Delineation of Male and Female Characters, Ancient and Modern, Who Have Been Particularly Distinguished by Extraordinary Qualifications, Talents, and Propensities, Natural or Acquired, Comprehending Singlar Instances of Longevity, Conformation, Bulk, Stature, Powers of Mind and of Body, Wonderful Exploits, Adventures, Habits, Propensities, Enterprising Pursuits, &C. &C. &C. With a Faithful Narration of Every Instance of Singularity Manifested in the Lives and Conduct of Characters Who Have Rendered Themselves Eminently Conspicuous by Their Eccentricities, the Whole Exhibiting an Interesting and Wonderful Display of Human Action in the Grand Theatre of the World. Collected and Re-Collected, from the Most Authentic Sources, by G. H. Wilson*, 4 vols (London: J. Cundee, 1806).

Wilson, R., and A. Mackley, *Creating Paradise: The Building of the English Country House 1660–1880* (London and New York: Hambledon, 2000).

Winter, A., *Mezmerised: Powers of Mind in Victorian Britain* (Chicago, IL: University of Chicago Press, 1998).

Wood, J., 'Modern Taxidermy', *Cornhill Magazine*, 1:7 (1863), pp. 120–5.

Wright, P., *On Living in an Old Country: The National Past in Contemporary Britain* (London: Verso, 1985).

Wright, T. G., *The Diary of Thomas Giordani Wright, Newcastle Doctor, 1826–1829* (Woodbridge: Boydell, 2001).

Yanni, C., *Nature's Museums: Victorian Science and the Architecture of Display* (London: The Athlone Press, 1999).

Yeo, R., *Defining Science: William Whewell, Natural Knowledge, and Pulbic Debate in Early Victorian Britain* (Cambridge: Cambridge University Press, 1993).

Yeo, R., *Encyclopaedic Visions: Scientific Dictionaries and Enlightenment Culture* (Cambridge: Cambridge University Press, 2001).

Yonge, C. M., *The Daisy Chain; Or, Aspirations. A Family Chronicle. By the Author of the Heir of Redclyffe, Etc.* (London: John W. Parker & Son, 1856). Nineteenth-Century Fiction Full-Text Database: Chadwyck-Healey, 2000.

Zonitch, B., *Familiar Violence: Gender and Social Upheaval in the Novels of Frances Burney* (London: Associated University Presses, 1997).

INDEX